ENVIRONMENTAL REMEDIATION TECHNOLOGIES, REGULATIONS AND SAFETY

NATIONAL STRATEGIES FOR THE ARCTIC AND A REVIEW OF ARCTIC CHANGES AND CONGRESSIONAL ISSUES

ENVIRONMENTAL REMEDIATION TECHNOLOGIES, REGULATIONS AND SAFETY

Additional books in this series can be found on Nova's website under the Series tab.

Additional E-books in this series can be found on Nova's website under the E-book tab.

Environmental Remediation Technologies, Regulations and Safety

National Strategies for the Arctic and a Review of Arctic Changes and Congressional Issues

Leonard DeLuca
Editor

nova publishers
New York

Library of Congress Cataloging-in-Publication Data

ISBN: 978-1-63117-201-4

Published by Nova Science Publishers, Inc. † New York

CONTENTS

PREFACE

This book discusses the National Strategy for the Arctic Region in further detail; the Department of Defense Arctic strategy; and provides background and issues in the changes in the Arctic.

Chapter 1 - The *National Strategy for the Arctic Region* sets forth the United States Government's strategic priorities for the Arctic region. This strategy is intended to position the United States to respond effectively to challenges and emerging opportunities arising from significant increases in Arctic activity due to the diminishment of sea ice and the emergence of a new Arctic environment. It defines U.S. national security interests in the Arctic region and identifies prioritized lines of effort, building upon existing initiatives by Federal, state, local, and tribal authorities, the private sector, and international partners, and aims to focus efforts where opportunities exist and action is needed. It is designed to meet the reality of a changing Arctic environment, while we simultaneously pursue our global objective of combating the climatic changes that are driving these environmental conditions. Our strategy is built on three lines of effort:

1. **Advance United States Security Interests** – We will enable our vessels and aircraft to operate, consistent with international law, through, under, and over the airspace and waters of the Arctic, support lawful commerce, achieve a greater awareness of activity in the region, and intelligently evolve our Arctic infrastructure and capabilities, including ice-capable platforms as needed. U.S. security in the Arctic encompasses a broad spectrum of activities, ranging from those supporting safe commercial and scientific operations to national defense.

2. **Pursue Responsible Arctic Region Stewardship** – We will continue to protect the Arctic environment and conserve its resources; establish and institutionalize an integrated Arctic management framework; chart the Arctic region; and employ scientific research and traditional knowledge to increase understanding of the Arctic.
3. **Strengthen International Cooperation** – Working through bilateral relationships and multilateral bodies, including the Arctic Council, we will pursue arrangements that advance collective interests, promote shared Arctic state prosperity, protect the Arctic environment, and enhance regional security, and we will work toward U.S. accession to the United Nations Convention on the Law of the Sea (Law of the Sea Convention).

Our approach will be informed by the following guiding principles:

- **Safeguard Peace and Stability** – Seek to maintain and preserve the Arctic region as an area free of conflict, acting in concert with allies, partners, and other interested parties. Support and preserve: international legal principles of freedom of navigation and overflight and other uses of the sea and airspace related to these freedoms, unimpeded lawful commerce, and the peaceful resolution of disputes for all nations.
- **Make Decisions Using the Best Available Information** – Across all lines of effort, decisions need to be based on the most current science and traditional knowledge.
- **Pursue Innovative Arrangements** – Foster partnerships with the State of Alaska, Arctic states, other international partners, and the private sector to more efficiently develop, resource, and manage capabilities, where appropriate and feasible, to better advance our strategic priorities in this austere fiscal environment.
- **Consult and Coordinate with Alaska Natives** – Engage in a consultation process with Alaska Natives, recognizing tribal governments' unique legal relationship with the United States and providing for meaningful and timely opportunity to inform Federal policy affecting Alaskan Native communities.

Chapter 2 - The Arcticis at a strategic inflection point as its ice cap is diminishing more rapidly than projected and human activity, driven by economic opportunity—ranging from oil, gas, and mineral exploration to fishing, shipping, and tourism—is increasing in response to the growing accessibility. Arctic and non-Arctic nations are establishing their strategies and positions on the future of the Arctic in a variety of international forums. Taken together, these changes present a compelling opportunity for the Department of Defense (DoD) to work collaboratively with allies and partners to promote a balanced approach to improving human and environmental security in the region in accordance with the 2013 *National Strategy for the Arctic Region.*

Security in the Arctic encompasses a broad spectrum of activities, ranging from resource extraction and trade to activities supporting safe commercial and scientific operations to national defense. Security cooperation activities and other military-to-military forms of engagement establish, shape, and maintain international relations and the partnerships necessary to meet security challenges and reduce the potential for friction. The Department will continue to build cooperative strategic partnerships that promote innovative, affordable security solutions, and burden-sharing in the Arctic, and seek to increase opportunities with Arctic partners to enhance regional expertise and cold-weather operational experience.

The Department will continue to train and operate routinely in the region as it monitors the changing environment, revisiting assessments and taking appropriate action as conditions change.

This strategy identifies the Department's desired end-state for the Arctic: ***a secure and stable region where U.S. national interests are safeguarded, the U.S. homeland is protected, and nations work cooperatively to address challenges***. It also articulates two main supporting objectives: *Ensure security, support safety, and promote defense cooperation*, and *prepare to respond to a wide range of challenges and contingencies*—operating in conjunction with other nations when possible, and independently if necessary—in order to maintain stability in the region. Finally, it identifies the *ways* and *means* the Department intends to use to achieve these objectives as it implements the *National Strategy for the Arctic Region.*

Chapter 3 - The diminishment of Arctic sea ice has led to increased human activities in the Arctic, and has heightened interest in, and concerns about, the region's future. The United States, by virtue of Alaska, is an Arctic country and has substantial interests in the region. On May 10, 2013, the Obama Administration released a national strategy document for the Arctic region.

Record low extents of Arctic sea ice over the past decade have focused scientific and policy attention on links to global climate change and projected ice-free seasons in the Arctic within decades. These changes have potential consequences for weather in the United States, access to mineral and biological resources in the Arctic, the economies and cultures of peoples in the region, and national security.

The five Arctic coastal states—the United States, Canada, Russia, Norway, and Denmark (of which Greenland is a territory)—are in the process of preparing Arctic territorial claims for submission to the Commission on the Limits of the Continental Shelf. The Russian claim to the enormous underwater Lomonosov Ridge, if accepted, would reportedly grant Russia nearly one- half of the Arctic area. There are also four other unresolved Arctic territorial disputes.

The diminishment of Arctic ice could lead in coming years to increased commercial shipping on two trans-Arctic sea routes—the Northern Sea Route and the Northwest Passage. Current international guidelines for ships operating in Arctic waters are being updated.

Changes to the Arctic brought about by warming temperatures will likely allow more exploration for oil, gas, and minerals. Warming that causes permafrost to melt could pose challenges to onshore exploration activities. Increased oil and gas exploration and tourism (cruise ships) in the Arctic increase the risk of pollution in the region. Cleaning up oil spills in ice-covered waters will be more difficult than in other areas, primarily because effective strategies have yet to be developed.

Large commercial fisheries exist in the Arctic. The United States is currently meeting with other countries regarding the management of Arctic fish stocks. Changes in the Arctic could affect threatened and endangered species. Under the Endangered Species Act, the polar bear was listed as threatened on May 15, 2008. Arctic climate change is also expected to affect the economies, health, and cultures of Arctic indigenous peoples.

Two of the Coast Guard's three polar icebreakers—*Polar Star* and *Polar Sea*—have exceeded their intended 30-year service lives, and *Polar Sea* is not operational. The possibility of increased sea traffic through Arctic waters raises an issue concerning Arctic search and rescue capabilities. On May 12, 2011, representatives from the member states of the Arctic Council signed an agreement on cooperation on aeronautical and maritime search and rescue in the Arctic.

The Arctic has increasingly become a subject of discussion among political leaders of the nations in the region. Although there is significant international cooperation on Arctic issues, the Arctic is also increasingly being viewed by some observers as a potential emerging security issue. In varying degrees, the Arctic coastal states have indicated a willingness to establish and maintain a military presence in the high north. U.S. military forces, particularly the Navy and Coast Guard, have begun to pay more attention to the region. On May 21, 2013, the Coast Guard released a strategy document for the Arctic.

In: National Strategies for the Arctic ... ISBN: 978-1-63117-201-4
Editors: Leonard Deluca

Chapter 1

NATIONAL STRATEGY FOR THE ARCTIC REGION*

Executive Office of the President

EXECUTIVE SUMMARY

"The United States is an Arctic Nation with broad and fundamental interests in the Arctic Region, where we seek to meet our national security needs, protect the environment, responsibly manage resources, account for indigenous communities, support scientific research, and strengthen international cooperation on a wide range of issues."[1]

The *National Strategy for the Arctic Region* sets forth the United States Government's strategic priorities for the Arctic region. This strategy is intended to position the United States to respond effectively to challenges and emerging opportunities arising from significant increases in Arctic activity due to the diminishment of sea ice and the emergence of a new Arctic environment. It defines U.S. national security interests in the Arctic region and identifies prioritized lines of effort, building upon existing initiatives by Federal, state, local, and tribal authorities, the private sector, and international partners, and aims to focus efforts where opportunities exist and action is needed. It is designed to meet the reality of a changing Arctic environment, while we simultaneously pursue our global objective of combating the climatic

---

* This is an edited, reformatted and augmented version of the Executive Office of the President publication, dated May 2013.

changes that are driving these environmental conditions. Our strategy is built on three lines of effort:

1. **Advance United States Security Interests** – We will enable our vessels and aircraft to operate, consistent with international law, through, under, and over the airspace and waters of the Arctic, support lawful commerce, achieve a greater awareness of activity in the region, and intelligently evolve our Arctic infrastructure and capabilities, including ice-capable platforms as needed. U.S. security in the Arctic encompasses a broad spectrum of activities, ranging from those supporting safe commercial and scientific operations to national defense.
2. **Pursue Responsible Arctic Region Stewardship** – We will continue to protect the Arctic environment and conserve its resources; establish and institutionalize an integrated Arctic management framework; chart the Arctic region; and employ scientific research and traditional knowledge to increase understanding of the Arctic.
3. **Strengthen International Cooperation** – Working through bilateral relationships and multilateral bodies, including the Arctic Council, we will pursue arrangements that advance collective interests, promote shared Arctic state prosperity, protect the Arctic environment, and enhance regional security, and we will work toward U.S. accession to the United Nations Convention on the Law of the Sea (Law of the Sea Convention).

Our approach will be informed by the following guiding principles:

- **Safeguard Peace and Stability** – Seek to maintain and preserve the Arctic region as an area free of conflict, acting in concert with allies, partners, and other interested parties. Support and preserve: international legal principles of freedom of navigation and overflight and other uses of the sea and airspace related to these freedoms, unimpeded lawful commerce, and the peaceful resolution of disputes for all nations.
- **Make Decisions Using the Best Available Information** – Across all lines of effort, decisions need to be based on the most current science and traditional knowledge.[2]

- **Pursue Innovative Arrangements** – Foster partnerships with the State of Alaska, Arctic states, other international partners, and the private sector to more efficiently develop, resource, and manage capabilities, where appropriate and feasible, to better advance our strategic priorities in this austere fiscal environment.
- **Consult and Coordinate with Alaska Natives** – Engage in a consultation process with Alaska Natives, recognizing tribal governments' unique legal relationship with the United States and providing for meaningful and timely opportunity to inform Federal policy affecting Alaskan Native communities.

INTRODUCTION

We seek an Arctic region that is stable and free of conflict, where nations act responsibly in a spirit of trust and cooperation, and where economic and energy resources are developed in a sustainable manner that also respects the fragile environment and the interests and cultures of indigenous peoples.

As the United States addresses these opportunities and challenges, we will be guided by our central interests in the Arctic region, which include providing for the security of the United States; protecting the free flow of resources and commerce; protecting the environment; addressing the needs of indigenous communities; and enabling scientific research. In protecting these interests, we draw from our long-standing policy and approach to the global maritime spaces in the 20th century, including freedom of navigation and overflight and other internationally lawful uses of the sea and airspace related to these freedoms; security on the oceans; maintaining strong relationships with allies and partners; and peaceful resolution of disputes without coercion.

To achieve this vision, the United States is establishing an overarching national approach to advance national security interests, pursue responsible stewardship of this precious and unique region, and serve as a basis for cooperation with other Arctic states[3] and the international community as a whole to advance common interests.

Even as we work domestically and internationally to minimize the effects of climate change, the effects are already apparent in the Arctic. Ocean resources are more readily accessible as sea ice diminishes, but thawing ground is threatening communities as well as hindering land-based activities, including access to resources.

Diminishing land and sea ice is altering ecosystems and the services they provide. As an Arctic nation, the United States must be proactive and disciplined in addressing changing regional conditions and in developing adaptive strategies to protect its interests. An undisciplined approach to exploring new opportunities in this frontier could result in significant harm to the region, to our national security interests, and to the global good.

When implementing this strategy, the United States will proceed in a thoughtful, responsible manner that leverages expertise, resources, and cooperation from the State of Alaska, Alaska Natives, and stakeholders across the entire nation and throughout the international community. We will encourage and use science-informed decisionmaking to aid this effort. We will endeavor to do no harm to the sensitive environment or to Alaska native communities and other indigenous populations that rely on Arctic resources. Just as a common spirit and shared vision of peaceful partnership led to the development of an international space station, we believe much can be achieved in the Arctic region through collaborative international efforts, coordinated investments, and public-private partnerships.

Structure of the Strategy

Through this *National Strategy for the Arctic Region,* we seek to guide, prioritize, and synchronize efforts to protect U.S. national and homeland security interests, promote responsible stewardship, and foster international cooperation.

This strategy articulates three priority lines of effort. It also identifies guiding principles as a foundation for Arctic region activities. Through a deliberate emphasis on the priority lines of effort and objectives, it aims to achieve a national unity of effort that is consistent with our domestic and international legal rights, obligations, and commitments and that is well coordinated with our Arctic neighbors and the international community. These lines of effort identify common themes where specific emphasis and activities will be focused to ensure that strategic priorities are met. The three lines of effort, as well as the guiding principles are meant to be acted upon as a coherent whole.

CHANGING CONDITIONS

While the Arctic region has experienced warming and cooling cycles over millennia, the current warming trend is unlike anything previously recorded. The reduction in sea ice has been dramatic, abrupt, and unrelenting. The dense, multi-year ice is giving way to thin layers of seasonal ice, making more of the region navigable year-round. Scientific estimates of technically recoverable conventional oil and gas resources north of the Arctic Circle total approximately 13 percent of the world's undiscovered oil and 30 percent of the world's undiscovered gas deposits, as well as vast quantities of mineral resources, including rare earth elements, iron ore, and nickel.[4] These estimates have inspired fresh ideas for commercial initiatives and infrastructure development in the region. As portions of the Arctic Ocean become more navigable, there is increasing interest in the viability of the Northern Sea Route and other potential routes, including the Northwest Passage, as well as in development of Arctic resources.

For all of the opportunities emerging with the increasing accessibility and economic and strategic interests in the Arctic, the opening and rapid development of the Arctic region presents very real challenges. On the environmental front, reduced sea ice is having an immediate impact on indigenous populations as well as on fish and wildlife. Moreover, there may be potentially profound environmental consequences of continued ocean warming and Arctic ice melt. These consequences include altering the climate of lower latitudes, risking the stability of Greenland's ice sheet, and accelerating the thawing of the Arctic permafrost in which large quantities of methane – a potent driver of climate change – as well as pollutants such as mercury are stored. Uncoordinated development – and the consequent increase in pollution such as emissions of black carbon or other substances from fossil fuel combustion – could have unintended consequences on climate trends, fragile ecosystems, and Arctic communities. It is imperative that the United States proactively establish national priorities and objectives for the Arctic region.

LINES OF EFFORT

To meet the challenges and opportunities in the Arctic region, and in furtherance of established Arctic Region Policy[5], we will pursue the following lines of effort and supporting objectives in a mutually reinforcing manner that incorporates the broad range of U.S. current activities and interests in the Arctic region.

1. Advance United States Security Interests

Our highest priority is to protect the American people, our sovereign territory and rights, natural resources, and interests of the United States. To this end, the United States will identify, develop, and maintain the capacity and capabilities necessary to promote safety, security, and stability in the region through a combination of independent action, bilateral initiatives, and multilateral cooperation. We acknowledge that the protection of our national security interests in the Arctic region must be undertaken with attention to environmental, cultural, and international considerations outlined throughout this strategy. As many nations across the world aspire to expand their role in the Arctic, we encourage Arctic and non-Arctic states to work collaboratively through appropriate fora to address the emerging challenges and opportunities in the Arctic region, while we remain vigilant to protect the security interests of the United States and our allies.

To accomplish this line of effort, the United States Government will seek to:

- **Evolve Arctic Infrastructure and Strategic Capabilities** – Working cooperatively with the State of Alaska, local, and tribal authorities, as well as public and private sector partners, we will develop, maintain, and exercise the capacity to execute Federal responsibilities in our Arctic waters, airspace, and coastal regions, including the capacity to respond to natural or man-made disasters. We will carefully tailor this regional infrastructure, as well as our response capacity, to the evolving human and commercial activity in the Arctic region.
- **Enhance Arctic Domain Awareness** – We seek to improve our awareness of activities, conditions, and trends in the Arctic region that may affect our safety, security, environmental, or commercial interests. The United States will endeavor to appropriately enhance sea, air, and space capabilities as Arctic conditions change, and to promote maritime-related information sharing with international, public, and private sector partners, to support implementation of activities such as the search-and-rescue agreement signed by Arctic states.

- **Preserve Arctic Region Freedom of the Seas** – The United States has a national interest in preserving all of the rights, freedoms, and uses of the sea and airspace recognized under international law. We will enable prosperity and safe transit by developing and maintaining sea, under-sea, and air assets and necessary infrastructure. In addition, the United States will support the enhancement of national defense, law enforcement, navigation safety, marine environment response, and search-and-rescue capabilities. Existing international law provides a comprehensive set of rules governing the rights, freedoms, and uses of the world's oceans and airspace, including the Arctic. The law recognizes these rights, freedoms, and uses for commercial and military vessels and aircraft. Within this framework, we shall further develop Arctic waterways management regimes, including traffic separation schemes, vessel tracking, and ship routing, in collaboration with partners. We will also encourage other nations to adhere to internationally accepted principles. This cooperation will facilitate strategic partnerships that promote innovative, low-cost solutions that enhance the Arctic marine transportation system and the safe, secure, efficient and free flow of trade.
- **Provide for Future United States Energy Security** – The Arctic region's energy resources factor into a core component of our national security strategy: energy security. The region holds sizable proved and potential oil and natural gas resources that will likely continue to provide valuable supplies to meet U.S. energy needs. Continuing to responsibly develop Arctic oil and gas resources aligns with the United States "all of the above" approach to developing new domestic energy sources, including renewables, expanding oil and gas production, and increasing efficiency and conservation efforts to reduce our reliance on imported oil and strengthen our nation's energy security. Within the context of this broader energy security strategy, including our economic, environmental and climate policy objectives, we are committed to working with stakeholders, industry, and other Arctic states to explore the energy resource base, develop and implement best practices, and share experiences to enable the environmentally responsible production of oil and natural gas as well as renewable energy.

2. Pursue Responsible Arctic Region Stewardship

Responsible stewardship requires active conservation of resources, balanced management, and the application of scientific and traditional knowledge of physical and living environments. As Arctic environments change, increased human activity demands precaution, as well as greater knowledge to inform responsible decisions. Together, Arctic nations can responsibly meet new demands – including maintaining open sea lanes for global commerce and scientific research, charting and mapping, providing search-and- rescue services, and developing capabilities to prevent, contain, and respond to oil spills and accidents – by increasing knowledge and integrating Arctic management.[6] We must improve our ability to forecast future conditions in the Arctic while being mindful of the potential for unexpected developments.

To realize this line of effort, we will pursue the specific objectives outlined below:

- **Protect the Arctic Environment and Conserve Arctic Natural Resources** – Protecting the unique and changing environment of the Arctic is a central goal of U.S. policy. Supporting actions will promote healthy, sustainable, and resilient ecosystems over the long term, supporting a full range of ecosystem services. This effort will be risk-based and proceed on the basis of best available information. The United States in the Arctic will assess and monitor the status of ecosystems and the risks of climate change and other stressors to prepare for and respond effectively to environmental challenges.
- **Use Integrated Arctic Management to Balance Economic Development, Environmental Protection, and Cultural Values** – Natural resource management will be based on a comprehensive understanding of environmental and cultural sensitivities in the region, and address expectations for future infrastructure needs and other development-related trends. This endeavor can promote unity of effort and provide the basis for sensible infrastructure and other resource management decisions in the Arctic. We will emphasize science-informed decisionmaking and integration of economic, environmental, and cultural values. We will also advance coordination among Federal departments and agencies and collaboration with partners engaged in Arctic stewardship activities.

- **Increase Understanding of the Arctic through Scientific Research and Traditional Knowledge** – Proper stewardship of the Arctic requires understanding of how the environment is changing, and such understanding will be based on a holistic earth system approach. Vast areas of the Arctic Ocean are unexplored, and we lack much of the basic knowledge necessary to understand and address Arctic issues. The changes in the Arctic cannot be understood in isolation and must be viewed in a global context. As we learn more about the region, we have identified several key subcomponents of the Arctic that require urgent attention: land ice and its role in changing sea level; sea-ice and its role in global climate, fostering biodiversity, and supporting Arctic peoples; and, the warming permafrost and its effects on infrastructure and climate. Better earth system-level knowledge will also help us meet operational needs such as weather and ice forecasting. We can make faster progress through a well-coordinated and transparent national and international exploration and research agenda that reduces the potential for duplication of effort and leads to better leveraging of resources.
- **Chart the Arctic region** – We will continue to make progress in charting and mapping the Arctic region's ocean and waterways, so long obscured by perennial ice, and mapping its coastal and interior lands according to reliable, modern standards. Given the vast expanse of territory and water to be charted and mapped, we will need to prioritize and synchronize charting efforts to make more effective use of resources and attain faster progress. In so doing, we will make navigation safer and contribute to the identification of ecologically sensitive areas and reserves of natural resources.

3. Strengthen International Cooperation

What happens in one part of the Arctic region can have significant implications for the interests of other Arctic states and the international community as a whole. The remote and complex operating conditions in the Arctic environment make the region well-suited for collaborative efforts by nations seeking to explore emerging opportunities while emphasizing ecological awareness and preservation. We will seek to strengthen partnerships through existing multilateral fora and legal frameworks dedicated to common Arctic issues. We will also pursue new arrangements for cooperating on issues of mutual interest or concern and addressing unique and unprecedented challenges, as appropriate.

U.S. efforts to strengthen international cooperation and partnerships will be pursued through four objectives:

- **Pursue Arrangements that Promote Shared Arctic State Prosperity, Protect the Arctic Environment, and Enhance Security** – We will seek opportunities to pursue efficient and effective joint ventures, based on shared values that leverage each Arctic state's strengths. This collaboration will assist in guiding investments and regional activities, addressing dynamic trends, and promoting sustainable development in the Arctic region.
  Arctic nations have varied commercial, cultural, environmental, safety, and security concerns in the Arctic region. Nevertheless, our common interests make these nations ideal partners in the region. We seek new opportunities to advance our interests by proactive engagement with other Arctic nations through bilateral and multilateral efforts using of a wide array of existing multilateral mechanisms that have responsibilities relating to the Arctic region.
  As appropriate, we will work with other Arctic nations to develop new coordination mechanisms to keep the Arctic region prosperous, environmentally sustainable, operationally safe, secure, and free of conflict, and will protect U.S., allied, and regional security and economic interests.
- **Work through the Arctic Council to Advance U.S. Interests in the Arctic Region** – In recent years, the Arctic Council has facilitated notable achievements in the promotion of cooperation, coordination, and interaction among Arctic states and Arctic indigenous peoples. Recent successes of the Council include its advancement of public safety and environmental protection issues, as evidenced by the 2011 Arctic Search-and-Rescue Agreement and by the 2013 Arctic Marine Oil Pollution Preparedness and Response Agreement. The United States will continue to emphasize the Arctic Council as a forum for facilitating Arctic states' cooperation on myriad issues of mutual interest within its current mandate.
- **Accede to the Law of the Sea Convention** – Accession to the Convention would protect U.S. rights, freedoms, and uses of the sea and airspace throughout the Arctic region, and strengthen our arguments for freedom of navigation and overflight through the

Northwest Passage and the Northern Sea Route. The United States is the only Arctic state that is not party to the Convention. Only by joining the Convention can we maximize legal certainty and best secure international recognition of our sovereign rights with respect to the U.S. extended continental shelf in the Arctic and elsewhere, which may hold vast oil, gas, and other resources. Our extended continental shelf claim in the Arctic region could extend more than 600 nautical miles from the north coast of Alaska.

- In instances where the maritime zones of coastal nations overlap, Arctic states have already begun the process of negotiating and concluding maritime boundary agreements, consistent with the Law of the Sea Convention and other relevant international law. The United States supports peaceful management and resolution of disputes, in a manner free from coercion. While the United States is not currently a party to the Convention, we will continue to support and observe principles of established customary international law reflected in the Convention.
- **Cooperate with other Interested Parties** – A growing number of non-Arctic states and numerous non-state actors have expressed increased interest in the Arctic region. The United States and other Arctic nations should seek to work with other states and entities to advance common objectives in the Arctic region in a manner that protects Arctic states' national interests and resources. One key example relates to the promotion of safe, secure, and reliable Arctic shipping, a goal that is best pursued through the International Maritime Organization in coordination with other Arctic states, major shipping states, the shipping industry and other relevant interests.

Guiding Principles

The U.S. approach to the Arctic region must reflect our values as a nation and as a member of the global community. We will approach holistically our interests in promoting safety and security, advancing economic and energy development, protecting the environment, addressing climate change and respecting the needs of indigenous communities and Arctic state interests. To guide our efforts, we have identified the following principles to serve as the foundation for U.S. Arctic engagement and activities.

- **Safeguard Peace and Stability** by working to maintain and preserve the Arctic region as an area free of conflict, acting in concert with allies, partners, and other interested parties. This principle will include United States action, and the actions of other interested countries, in supporting and preserving international legal principles of freedom of navigation and overflight and other uses of the sea related to these freedoms, unimpeded lawful commerce, and the peaceful resolution of disputes. The United States will rely on existing international law, which provides a comprehensive set of rules governing the rights, freedoms, and uses of the world's oceans and airspace, including the Arctic.
- **Make Decisions Using the Best Available Information** by promptly sharing – nationally and internationally – the most current understanding and forecasts based on up-to-date science and traditional knowledge.
- **Pursue Innovative Arrangements** to support the investments in scientific research, marine transportation infrastructure requirements, and other support capability and capacity needs in this region. The harshness of the Arctic climate and the complexity associated with developing, maintaining, and operating infrastructure and capabilities in the region necessitate new thinking on public-private and multinational partnerships.
- **Consult and Coordinate with Alaska Natives** consistent with tribal consultation policy established by Executive Order.[7] This policy emphasizes trust, respect, and shared responsibility. It articulates that tribal governments have a unique legal relationship with the United States and requires Federal departments and agencies to provide for meaningful and timely input by tribal officials in development of regulatory policies that have tribal implications. This guiding principle is also consistent with the Alaska Federation of Natives Guidelines for Research.[8]

CONCLUSION

We seek a collaborative and innovative approach to manage a rapidly changing region. We must advance U.S. national security interests, pursue responsible stewardship, and strengthen international collaboration and cooperation, as we work to meet the challenges of rapid climate-driven environmental change. The melting of Arctic ice has the potential to transform global climate and ecosystems as well as global shipping, energy markets, and other commercial interests. To address these challenges and opportunities, we will align Federal activities in accordance with this strategy; partner with the State of Alaska, local, and tribal entities; and work with other Arctic nations to develop complementary approaches to shared challenges. We will proactively coordinate regional development. Our economic development and environmental stewardship must go hand-in-hand The unique Arctic environment will require a commitment by the United States to make judicious, coordinated infrastructure investment decisions, informed by science. To meet this challenge, we will need bold, innovative thinking that embraces and generates new and creative public-private and multinational cooperative models.

End Notes

[1] National Security Strategy, May 2010.

[2] Traditional knowledge refers to a body of evolving practical knowledge based on observations and personal experience of indigenous communities over an extensive, multigenerational time period. (BOEM Ocean Science, Vol. 9, Issue 2, May/April/June 2012, page 4).

[3] Arctic state is defined as one of the eight nations making up the permanent membership of the Arctic Council and includes the following nations: Canada, Denmark (including Greenland and the Faroe Islands), Finland, Iceland, Norway, Russia, Sweden, and the United States.

[4] U.S. Geological Survey, "Circum-Arctic Resource Appraisal: Estimates of Undiscovered Oil and Gas North of the Arctic Circle," fact sheet, 2008, http://pubs.usgs.gov/fs/2008/3049/fs2008-3049.pdf.

[5] National Security Presidential Directive-66/Homeland Security Presidential Directive-25, January 2009.

[6] Much of this work is already underway including efforts under Executive Order 12501 (Arctic Research), Executive Order 13547 (Stewardship of the Ocean, Our Coasts, and the Great Lakes), and Executive Order 13580 (Interagency Working Group on Coordination of Domestic Energy Development and Permitting in Alaska). Entities under these Executive Orders are developing partnerships with Federal, state, local, tribal, territorial, public and private sector partners to ensure that natural resource decisions in the Arctic integrate economic, environmental, and cultural interests of the Nation.

[7] See Executive Order 13175 – Consultation and Coordination with Indian Tribal Governments, November 2000.

[8] Alaska Federation of Natives (AFN) (1996) *AFN Board Adopts Policy Guidelines for Research.* AFN, Anchorage. Alaska Natives Knowledge Network Policy adopted in 1993, http://ankn.uaf.edu/IKS/afnguide.html.

In: National Strategies for the Arctic ... ISBN: 978-1-63117-201-4
Editors: Leonard Deluca

Chapter 2

DEPARTMENT OF DEFENSE ARCTIC STRATEGY*

EXECUTIVE SUMMARY

The Arctic[1] is at a strategic inflection point as its ice cap is diminishing more rapidly than projected[2] and human activity, driven by economic opportunity—ranging from oil, gas, and mineral exploration to fishing, shipping, and tourism—is increasing in response to the growing accessibility. Arctic and non-Arctic nations are establishing their strategies and positions on the future of the Arctic in a variety of international forums. Taken together, these changes present a compelling opportunity for the Department of Defense (DoD) to work collaboratively with allies and partners to promote a balanced approach to improving human and environmental security in the region in accordance with the 2013 *National Strategy for the Arctic Region*.[3]

Security in the Arctic encompasses a broad spectrum of activities, ranging from resource extraction and trade to activities supporting safe commercial and scientific operations to national defense. Security cooperation activities and other military-to-military forms of engagement establish, shape, and maintain international relations and the partnerships necessary to meet security challenges and reduce the potential for friction. The Department will continue to build cooperative strategic partnerships that promote innovative, affordable security solutions, and burden-sharing in the Arctic, and seek to increase

* This is an edited, reformatted and augmented version of a Department of Defense publication, dated November 2013.

opportunities with Arctic partners to enhance regional expertise and cold-weather operational experience.

The Department will continue to train and operate routinely in the region[4] as it monitors the changing environment, revisiting assessments and taking appropriate action as conditions change.

This strategy identifies the Department's desired end-state for the Arctic: ***a secure and stable region where U.S. national interests are safeguarded, the U.S. homeland is protected, and nations work cooperatively to address challenges***. It also articulates two main supporting objectives: *Ensure security, support safety, and promote defense cooperation*, and *prepare to respond to a wide range of challenges and contingencies*—operating in conjunction with other nations when possible, and independently if necessary—in order to maintain stability in the region. Finally, it identifies the *ways* and *means* the Department intends to use to achieve these objectives as it implements the *National Strategy for the Arctic Region.*

An iceberg at the edge of Baffin Bay's sea ice is just one of the many sights from Thule Air Base, Greenland. (U.S. Air Force photo by Tech. Sgt. Dan Rea).

I. U.S. Interests in the Arctic

U.S. national security interests in the Arctic are delineated in National Security Presidential Directive (NSPD) 66/Homeland Security Presidential Directive (HSPD) 25, *Arctic Region Policy.* [5] This policy states that national security interests include such matters as missile defense and early warning; deployment of sea and air systems for strategic sealift, strategic deterrence, maritime presence, and maritime security operations; and ensuring freedom of the seas. Preserving freedom of the seas, which includes all of the rights, freedoms, and uses of the seas and adjacent airspace, including freedom of navigation and overflight, in the Arctic supports the nation's ability to exercise these rights, freedoms, and uses of the sea and airspace throughout the world, including through strategic straits.

Arctic Boundary as defined by the Arctic Research and Policy Act (ARPA)

All United States and foreign territory north of the Arctic Circle and all United States territory north and west of the boundary formed by the Porcupine, Yukon, and Kuskokwim Rivers; all contiguous seas, including the Arctic Ocean and the Beaufort, Bering and Chukchi Seas, and the Aleutian chain.[1]

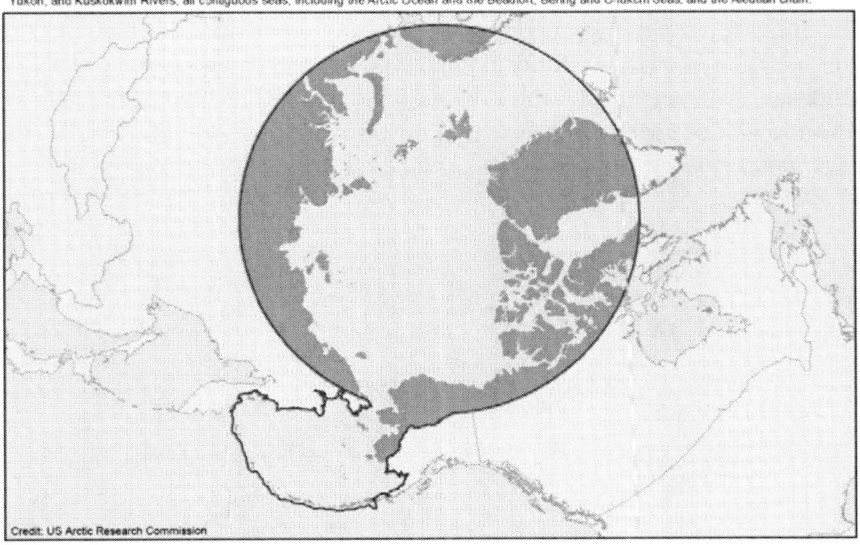

Acknowledgement: Funding for this map was provided by the National Science Foundation through the Arctic Research Mapping Application (armap.org) and Contract #0520837 to CH2M HILL for the Interagency Arctic Research Policy Committee (IARPC).

Map author: Allison Gaylord, Nuna Technologies, May 27, 2009.

1. The Aleutian chain boundary is demarcated by the 'Contiguous zone' limit of 24-nautical miles.

The 2013 *National Strategy on the Arctic Region* frames the whole-of-government approach that provides the overarching context for the Department's efforts. It lays out three main lines of effort in the Arctic: advance U.S. security interests; pursue responsible Arctic region stewardship; and strengthen international cooperation. The goal of the *National Strategy for the Arctic Region* is "an Arctic region that is stable and free of conflict, where nations act responsibly in a spirit of trust and cooperation, and where economic and energy resources are developed in a sustainable manner that also respects the fragile environment and the interests and cultures of indigenous peoples."

The DoD Arctic Strategy outlines how the Department will support the whole-of-government effort to promote security, stewardship, and international cooperation in the Arctic. The Department's strategic approach to the Arctic reflects the relatively low level of military threat in a region bounded by nation States that have not only publicly committed to working within a common framework of international law and diplomatic engagement,[6] but have also demonstrated the ability and commitment to do so. In consideration of enduring national interests in the Arctic and existing strategic guidance, the Department's end-state for its strategic approach to the Arctic is: ***a secure and stable region where U.S. national interests are safeguarded, the U.S. homeland is protected, and nations work cooperatively to address challenges***.

Members of the Applied Physics Laboratory Ice Station participate in Ice Exercise (ICEX) to assist operations and training in the challenging and unique environment that characterizes the Arctic region. (U.S. Navy photo by Mass Communication Specialist 1st Class Tiffini M. Jones).

II. Department of Defense Supporting Objectives

The Department's two supporting objectives describe what is to be accomplished to achieve its desired end-state. These objectives are bounded by policy guidance, the changing nature of the strategic and physical environment, and the capabilities and limitations of the available instruments of national power (diplomatic, informational, military, and economic). Actions taken to achieve these objectives will be informed by the Department's global priorities and fiscal constraints. In order to achieve its strategic end- state, the Department's supporting objectives are:

- ***Ensure security, support safety, and promote defense cooperation***.[7]
  - Relationships with allies and partners are important enablers of cooperation in meeting security and defense commitments. These relationships also play an important role in conflict prevention, and, if prevention and deterrence fail, in coordinating an international response to security and defense challenges. Although the Department of State is the lead for regional diplomacy, DoD has a supporting role enhancing the region's capability and capacity for multilateral security collaboration, and responding to requests for assistance from interagency and international partners both within and outside the Arctic. This collaborative approach helps prevent conflict and provides the stability needed to facilitate the sustainable economic development envisioned in the *National Strategy for the Arctic Region*. The Department of Defense will seek out areas of mutual interest to build strategic relationships and encourage operational-level partnerships that promote innovative, affordable security solutions and enhance burden- sharing in the Arctic. Science and technology (S&T) can provide non-contentious opportunities for cooperation, and DoD will coordinate research initiatives with the Interagency Arctic Research Policy Committee (IARPC).
  - The Department has an important role supporting other Federal departments and agencies in safety-related missions in Alaska and in responding to requests from civil authorities to support them with disaster relief or humanitarian assistance at home or abroad. Although the Department has seldom been tasked to execute these missions in the Arctic, it may be asked to do more in the coming decades.

- ***Prepare for a wide range of challenges and contingencies***—operating in conjunction with other States when possible and independently if necessary—in order to maintain stability in the region.
  - Future challenges in the Arctic may span the full range of national security interests. These challenges and contingencies may take many forms, ranging from the need to support other Federal departments and agencies—or another nation—in responding to a natural or man-made disaster to responding to security concerns that may emerge in the future.

III. Strategic Approach

The Department will pursue comprehensive engagement with allies and partners to protect the homeland and support civil authorities in preparing for increased human activity in the Arctic. Strategic partnerships are the center of gravity in ensuring a peaceful opening of the Arctic and achieving the Department's desired end-state. Where possible, DoD will seek innovative, low-cost, small- footprint approaches to achieve these objectives (e.g., by participating in multilateral exercises like the Search and Rescue Exercise (SAREX) hosted by Greenland, COLD RESPONSE hosted by Norway, and Canada's Operation NANOOK, or through Defense Environmental International Cooperation Program-supported engagements on Arctic issues). The Department will also evolve its infrastructure and capabilities in step with the changing physical environment in order to ensure security, support safety, promote defense cooperation, and prepare to respond to a wide range of challenges and contingencies in the Arctic in the coming decades. The Department will accomplish its objectives through the following *ways*:

- Exercise sovereignty and protect the homeland;
- Engage public and private sector partners to improve domain awareness in the Arctic;
- Preserve freedom of the seas in the Arctic;
- Evolve Arctic infrastructure and capabilities consistent with changing conditions;
- Support existing agreements with allies and partners while pursuing new ones to build confidence with key regional partners;

Soldiers conduct medical evacuation training at Fort Wainwright, Alaska. (USARPAC Public Affairs photo).

- Provide support to civil authorities, as directed;
- Partner with other departments and agencies and nations to support human and environmental safety; and
- Support the development of the Arctic Council and other international institutions that promote regional cooperation and the rule of law. [8]

The Department will apply the four guiding principles from the 2013 *National Strategy for the Arctic Region* as it pursues these eight *ways*.[9] This means DoD will work with allies, partners, and other interested parties to safeguard peace and stability. It will make decisions using the best available scientific information, and will pursue innovative arrangements as it develops the capability and capacity needed in the Arctic over time. It will also follow established Federal tribal consultation policy. These four principles will underpin all of the Department's activities as it implements this strategy through the *means* described in this section.

Protect the Homeland and Exercise Sovereignty. From the U.S. perspective, greater access afforded by the decreasing seasonal ice increases the Arctic's viability as an avenue of approach to North America for those with hostile intent toward the U.S. homeland, and the Department will remain prepared to detect, deter, prevent, and defeat threats to the homeland. Additionally, DoD will continue to support the exercise of U.S. sovereignty. In the near-term[10], this will require some ability to operate in the Arctic, which the Department will maintain and enhance by continuing to conduct exercises and training in the region. In the mid- to far-term, this may require developing further capabilities and capacity to protect U.S. air, land, and maritime borders in the Arctic in accordance with the 2013 *National Strategy for the Arctic Region.*

As directed by the 2011 Unified Command Plan, Commander, U.S. Northern Command (CDRUSNORTHCOM) is responsible for advocating for Arctic capabilities. In execution of this responsibility, CDRUSNORTHCOM will collaborate with relevant Combatant Commands, the Joint Staff, the Military Departments and Services, and the Defense agencies to identify and prioritize emerging Arctic capability gaps and requirements. These efforts will be informed by the most authoritative scientific information on future Arctic conditions. For purposes of mission and infrastructure vulnerability assessments and adaptation to climate change, the Office of the Under Secretary of Defense for Acquisition, Technology, and Logistics (OUSD(AT&L)) will identify projections of future conditions to be used. The Department of Defense will collaborate with the Department of Homeland Security (DHS) to ensure efficient use of resources to avoid duplication of effort in research, development, experimentation, testing, and acquisition. Forums such as the DoD-DHS Capabilities Development Working Group are among the means to facilitate this cooperation.

A C-17 Globemaster III aircraft and its aircrew assist in deploying 120 Soldiers for Operation Arctic Response, an Army Emergency Deployment Readiness Exercise. (U.S. Air Force photo by Senior Airman Laura Turner).

Engage public and private sector partners to improve all domain awareness in the Arctic. Although NSPD-66/HSPD-25 focuses on maritime domain awareness, the Department has responsibilities for awareness across all domains: air, land, maritime, space, and cyberspace. Adequate domain awareness is an essential component of protecting maritime commerce, critical infrastructure, and key resources. In the near-term, the Department will work through the North American Aerospace Defense Command (NORAD) to maintain air tracking capabilities in the Arctic. As the maritime domain becomes increasingly accessible, the Department will seek to improve its maritime detection and tracking in coordination with DHS and other departments and agencies as well as through public/private partnerships. The Department of the Navy, in its role as DoD Executive Agent for Maritime Domain Awareness, will lead DoD coordination on maritime detection and tracking. Where possible, DoD will also collaborate with international partners to employ, acquire, share, or develop the means required to improve sensing, data collection and fusion, analysis, and information-sharing to enhance domain awareness appropriately in the Arctic. Monitoring regional activity and analyzing emerging trends are key to informing future investments in Arctic capabilities and ensuring they keep pace with increasing human activity in the region over time.

Scientists from the Naval Undersea Warfare Center monitor icebergs while a topside computer system gathers data collected from an autonomous undersea vehicle. (U.S. Navy photo).

In the near- to mid-term, the primary means of improving domain awareness will be continued use of innovative, low-cost solutions for polar Command, Control, Communications, Computers, Intelligence, Surveillance, and Reconnaissance (C4ISR) needs as well as enhanced international collaboration. DoD will take steps to work with other Federal departments and agencies to improve nautical charts, enhance relevant atmospheric and oceanic models, improve accuracy of estimates of ice extent and thickness, and detect and monitor climate change indicators. In particular, the Department of the Navy will work in partnership with other Federal departments and agencies (e.g., DHS, the Department of Commerce) and international partners to improve hydrographic charting and oceanographic surveys in the Arctic.

The Department will continue to collaborate with other Federal departments and agencies and the State of Alaska to monitor and assess changes in the physical environment to inform the development of Arctic requirements and future capabilities. To that end, the Department will leverage work done by the scientific and academic communities and seek opportunities to contribute to the observation and modeling of the atmosphere, ocean, and sea ice conditions, including acoustics conditions, to enhance military environmental forecasting capabilities. These collaborations will help inform the development and design of future ice-strengthened ship designs, when required.

Preserve freedom of the seas in the Arctic. The United States has a national interest in preserving all of the rights, freedoms, and uses of the sea and airspace recognized under international law. The Department will preserve the global mobility of United States military and civilian vessels and aircraft throughout the Arctic,[11] including through the exercise of the Freedom of Navigation program to challenge excessive maritime claims asserted by other Arctic States when necessary. The Department will continue to support U.S. accession to the United Nations Convention on the Law of the Sea (hereafter referred to as the Law of the Sea Convention (LOSC)) because it codifies the rights, freedoms, and uses of the sea and airspace the Department seeks to preserve; provides a means for the peaceful resolution of disputes; and ensures international recognition of resources rights on the extended continental shelf.

Evolve Arctic infrastructure and capabilities consistent with changing conditions. The Department will periodically re-evaluate requirements necessary to meet national security objectives as conditions change and the Combatant Commanders identify operational requirements for the Arctic in updates to their regional plans. Once operational requirements are defined, solutions for associated supporting infrastructure requirements should seek to leverage existing U.S. Government, commercial, and international facilities to the maximum extent possible in order to mitigate the high cost and extended timelines associated with the development of Arctic infrastructure. If no existing infrastructure is capable of sufficiently supporting the requirement, modifications to existing bases, such as the addition of a new hangar, will be made as part of the military construction or facilities sustainment, restoration, and modernization processes.

Uphold existing agreements with allies and partners while building confidence with key regional partners. Security cooperation activities and other military-to-military forms of engagement establish, shape, and maintain international relations and the partnerships necessary to meet security challenges and reduce the potential for friction. The 2012 and 2013 Northern Chiefs of Defense (CHoDs) meetings and the Arctic Security Forces Roundtable workshops and meetings are examples of means for promoting information-sharing and partnership-building necessary to develop cooperative approaches to common challenges. Therefore, in cooperation with the Department of State, DHS (in particular, the U.S. Coast Guard), and other relevant agencies, the Department will continue to build cooperative strategic partnerships that promote innovative, affordable security solutions and burden-sharing in the Arctic. It will also seek to increase bilateral exchanges, including in science and technology, and take advantage of multilateral training opportunities with Arctic partners to enhance regional expertise and cold-weather operational experience.

Provide Defense Support of Civil Authorities (DSCA) in Alaska and provide Foreign Humanitarian Assistance and Foreign Disaster Relief (FHA/FDR) in other non-U.S. territorial areas of the Arctic. When directed by the appropriate authority, the Department will be prepared to support civil authorities in response to natural or manmade disasters, or to conduct FHA/FDR operations in cooperation with allies and partners.

Partner with other agencies and nations to support human and environmental safety. Some of the near-term safety-related challenges include meeting international search and rescue obligations and responding to incidents such as oil spills in ice-covered waters, as reflected in the recently concluded Agreement on Cooperation on Aeronautical and Maritime Search and Rescue in the Arctic and Agreement on Cooperation on Marine Oil Pollution Preparedness and Response in the Arctic. The Department will leverage existing capabilities to respond to requests for support from civil authorities in coordination with other departments and agencies and nations. Where appropriate, the Department will support other departments and agencies in maintaining human health; promoting healthy, sustainable, and resilient ecosystems; and consulting and coordinating with Alaska Natives on policy and activities affecting them. Finally, the Department will continue to integrate environmental considerations into its planning and operations and to contribute to whole-of-government approaches in support of the second line of effort in the 2013 *National Strategy for the Arctic Region.*

Support the development of the Arctic Council and other international institutions to promote regional cooperation and the rule of law. The Department recognizes the value of the Arctic Council in efforts to understand the changing Arctic environment and developing cooperative approaches to regional challenges, and supports the Department of State in the continued development of the Council. Although the Department of State is the lead for diplomacy, DoD has a role to play in enhancing the region's multilateral security cooperation environment. Accordingly, DoD will work with allies and partners within the framework of international institutions, ranging from the Arctic Council to the International Maritime Organization (IMO), to maintain stability and promote cooperation.

From South Mountain, Thule Air Base and Mount Dundas are captured during a short period of January twilight in Greenland. (U.S. Air Force photo by Michael Tolzmann).

IV. CHALLENGES & RISKS TO THE STRATEGIC APPROACH

This strategy furthers defense objectives while positioning the United States to take advantage of opportunities in the Arctic during the coming decades in accordance with the 2013 *National Strategy for the Arctic Region*. It also addresses some of the risks inherent in the trade-offs and tensions among U.S. interests and objectives, including:

- *Projections about future access to and activity in the Arctic may be inaccurate.* Significant uncertainty remains about the rate and extent of the effects of climate change, including climate variability, in the Arctic. There is also uncertainty about future economic conditions, and the pace at which human activity will increase in the region. The challenge is to balance the risk of having inadequate capabilities or insufficient capacity when required to operate in the region with the opportunity cost of making premature and/or unnecessary investments. Premature investment may reduce the availability of resources for other pressing priorities, particularly in a time of fiscal austerity. The key will be to address needs in step with the rate at which activity in the Arctic increases while balancing potential investments in Arctic capabilities with other national priorities. The Department will mitigate this risk by monitoring the changing Arctic environment and geostrategic situation to determine the appropriate timing for future capability investments. Designating CDRUSNORTHCOM as the Arctic capability advocate was a significant step forward in the effort, but success will require Departmental adaptability in planning, budgeting, and acquisitions.
- *Fiscal constraints may delay or deny needed investment in Arctic capabilities, and may curtail Arctic training and operations.* As the Department downsizes to meet budgetary targets, it will have to prioritize engagements for the resulting smaller force. There is also a risk that desired investments in Arctic capabilities may not compete successfully against other requirements in the Department's budgetary priorities. Where possible, DoD will mitigate this risk by developing innovative ways to employ existing capabilities in coordination with other departments and agencies and international partners, and by enhancing scientific, research, and development partnerships.

CDRUSNORTHCOM plays a key role in mitigating this risk as the Arctic capability advocate within the Department's planning and program-ing activities. Commander, U.S. European Command (CDRUSEUCOM) and Commander, U.S. Pacific Command (CDRUSPACOM) also play a role by fostering collaborative working relationships with regional partners.

- *Political rhetoric and press reporting about boundary disputes and competition for resources may inflame regional tensions.* Efforts to manage disagreements diplomatically may be hindered if the public narrative becomes one of rivalry and conflict. The Department will mitigate this risk by ensuring its plans, actions, and words are coordinated, and when appropriate, by engaging the press to counter unhelpful narratives with facts. The Under Secretary of Defense for Policy will monitor DoD activities, programs, and posture in the region to ensure the Department is sending a clear message to key audiences regarding the Department's efforts to promote security, safety, and defense cooperation.
- *Being too aggressive in taking steps to address anticipated future security risks may create the conditions of mistrust and miscommunication under which such risks could materialize.* There is some risk that the perception that the Arctic is being militarized may lead to an "arms race" mentality that could lead to a breakdown of existing cooperative approaches to shared challenges. The Department will mitigate this risk by focusing on collaborative security approaches as outlined in the 2013 *National Strategy for the Arctic Region*, and by supporting other Federal departments and agencies where they have leadership roles. Building trust through transparency about the intent of our military activities and participation in bilateral and multilateral exercises and other engagements that facilitate information-sharing will be a key means of addressing this risk.

CONCLUSION

The Department will work collaboratively with allies and partners through the ways and means outlined in this strategy to support the development of the Arctic as a secure and stable region where U.S. national interests are safeguarded, the U.S. homeland is protected, and nations work cooperatively to address challenges. Priority will be given to addressing key near-term challenges primarily in key enablers, including: shortfalls in ice and weather reporting and forecasting; limitations in C4ISR due to lack of assets and harsh environmental conditions; and limited domain awareness. The key will be to address needs over time as activity in the Arctic increases, while balancing potential Arctic investments with other national priorities. This approach will help the United States achieve its objectives as outlined in the *National Strategy for the Arctic Region* while mitigating risks and overcoming challenges presented by the growing geostrategic importance of the Arctic.

Arctic sea ice. (NOAA photo).

Arctic Survival School teaches signal fires. (U.S. Air Force photo by Airman 1st Class Willard E. Grande II).

ARCTIC REGION

The world Factbook 2013-14. Washington, DC: Central Intelligence Agency, 2013.

End Notes

[1] The DoD strategy uses a broad definition of the Arctic, codified in 15 U.S.C. 4111, that includes all U.S. and foreign territory north of the Arctic Circle and all U.S. territory north and west of the boundary formed by the Porcupine, Yukon, and Kuskokwim Rivers; all contiguous seas, including the Arctic Ocean and the Beaufort, Bering, and Chukchi Seas, and the Aleutian islands chain.

[2] National Oceanic and Atmospheric Administration (NOAA) scientists in J.E. Overland and M. Wang (2013), *When will the summer Arctic be nearly sea ice free?*, Geophys. Res. Lett., 40, doi:10.1002/grl.50316.

[3] This strategy is nested under *National Security Presidential Directive (NSPD) 66 / Homeland Security Presidential Directive (HSPD) 25,* Arctic Region Policy, the 2013 *National Strategy for the Arctic Region*, and the 2012 Defense Strategic Guidance. It complements DoD's *Strategy for Homeland Defense and Defense Support of Civil Authorities (HD&DSCA).*

[4] For additional information on the Navy's historic involvement in the Arctic, see *The Impact of Climate Change on Naval Operations in the Arctic* (Center for Naval Analysis, 2009).

[5] The January 2009 National Security Presidential Directive (NSPD)-66, dual-titled as Homeland Security Presidential Directive (HSPD)-25, or NSPD-66/HSPD-25, established the policy of the United States with respect to the Arctic and outlined national security and homeland security interests in the region. Homeland security interests include preventing terrorist attacks and mitigating those criminal or hostile acts that could increase the United States' vulnerability to terrorism in the Arctic. The Department has a role to play in responding not only to traditional (e.g., military) threats, but also to a range of other potential national security challenges (e.g., smuggling, criminal trafficking, and terrorism as the lead agency or in support of other government agencies).

[6] In the Ilulissat Declaration (May 28, 2008), all five Arctic Ocean coastal States (United States, Russian Federation, Canada, Norway, and Denmark on behalf of Greenland) committed themselves to the orderly settlement of overlapping territorial claims through the established framework of the international law as reflected in the U.N. Law of the Sea Convention (LOSC). The Declaration affirmed that the legal framework provided by the LOSC is sufficient for the management of the Arctic Ocean and that there is no need to develop a new comprehensive international legal regime to govern this Ocean.

[7] Per the 2013 *National Strategy for the Arctic Region*, U.S. security in the Arctic encompasses a broad spectrum of activities, including national defense.

[8] The Arctic Council's charter states, "The Arctic Council should not deal with matters related to military security." It could be argued that search and rescue is a (human) security interest, and oil spill response is an (environmental) security interest; thus, the Council has a demonstrated ability to address a range of "soft security" issues.

[9] Many of DoD's *ways* align with what the National Strategy for the Arctic Region terms *supporting objectives* to its three lines of effort, but DoD's strategy follows the classical "ways-ends-means" construction.

[10] This strategy identifies three timeframes to be used for implementation planning: the near-term (present day-2020); mid-term (2020-2030); and far-term (beyond 2030). These timeframes are approximate due to uncertainty in climate change projections.

[11] As expressed by Commander, U.S. Pacific Command (USPACOM), Commander, U.S. Transportation Command (USTRANSCOM), and Commander, USNORTHCOM, in a May 2008 memorandum, the United States needs assured access to support U.S. national interests in the Arctic. Although this imperative could be met by regular U.S. Government ships in open water up to the marginal ice zone, only ice-capable ships provide assured sovereign presence throughout the region and throughout the year. Assured access in areas of pack ice could also be met by other means, including submarines and aircraft.

In: National Strategies for the Arctic ... ISBN: 978-1-63117-201-4
Editors: Leonard Deluca

Chapter 3

CHANGES IN THE ARCTIC: BACKGROUND AND ISSUES FOR CONGRESS*

Ronald O'Rourke

SUMMARY

The diminishment of Arctic sea ice has led to increased human activities in the Arctic, and has heightened interest in, and concerns about, the region's future. The United States, by virtue of Alaska, is an Arctic country and has substantial interests in the region. On May 10, 2013, the Obama Administration released a national strategy document for the Arctic region.

Record low extents of Arctic sea ice over the past decade have focused scientific and policy attention on links to global climate change and projected ice-free seasons in the Arctic within decades. These changes have potential consequences for weather in the United States, access to mineral and biological resources in the Arctic, the economies and cultures of peoples in the region, and national security.

The five Arctic coastal states—the United States, Canada, Russia, Norway, and Denmark (of which Greenland is a territory)—are in the process of preparing Arctic territorial claims for submission to the Commission on the Limits of the Continental Shelf. The Russian claim to the enormous underwater Lomonosov Ridge, if accepted, would reportedly grant Russia

* This is an edited, reformatted and augmented version of a Congressional Research Service publication, CRS Report for Congress R41153, dated August 8, 2013.

nearly one- half of the Arctic area. There are also four other unresolved Arctic territorial disputes.

The diminishment of Arctic ice could lead in coming years to increased commercial shipping on two trans-Arctic sea routes—the Northern Sea Route and the Northwest Passage. Current international guidelines for ships operating in Arctic waters are being updated.

Changes to the Arctic brought about by warming temperatures will likely allow more exploration for oil, gas, and minerals. Warming that causes permafrost to melt could pose challenges to onshore exploration activities. Increased oil and gas exploration and tourism (cruise ships) in the Arctic increase the risk of pollution in the region. Cleaning up oil spills in ice-covered waters will be more difficult than in other areas, primarily because effective strategies have yet to be developed.

Large commercial fisheries exist in the Arctic. The United States is currently meeting with other countries regarding the management of Arctic fish stocks. Changes in the Arctic could affect threatened and endangered species. Under the Endangered Species Act, the polar bear was listed as threatened on May 15, 2008. Arctic climate change is also expected to affect the economies, health, and cultures of Arctic indigenous peoples.

Two of the Coast Guard's three polar icebreakers—*Polar Star* and *Polar Sea*—have exceeded their intended 30-year service lives, and *Polar Sea* is not operational. The possibility of increased sea traffic through Arctic waters raises an issue concerning Arctic search and rescue capabilities. On May 12, 2011, representatives from the member states of the Arctic Council signed an agreement on cooperation on aeronautical and maritime search and rescue in the Arctic.

The Arctic has increasingly become a subject of discussion among political leaders of the nations in the region. Although there is significant international cooperation on Arctic issues, the Arctic is also increasingly being viewed by some observers as a potential emerging security issue. In varying degrees, the Arctic coastal states have indicated a willingness to establish and maintain a military presence in the high north. U.S. military forces, particularly the Navy and Coast Guard, have begun to pay more attention to the region. On May 21, 2013, the Coast Guard released a strategy document for the Arctic.

INTRODUCTION

The diminishment of Arctic sea ice has led to increased human activities in the Arctic, and has heightened interest in, and concerns about, the region's future. Issues such as Arctic sovereignty claims; commercial shipping through the Arctic; Arctic oil, gas, and mineral exploration; endangered Arctic species; and increased military operations in the Arctic could cause the region in coming years to become an arena of international cooperation, competition, or conflict. The United States, by virtue of Alaska, is an Arctic country and has substantial political, economic, energy, environmental, and other interests in the region. Decisions that Congress, the executive branch, foreign governments, international organizations, and commercial firms make on Arctic- related issues could significantly affect these interests.

BACKGROUND[1]

Definitions of the Arctic

There are multiple definitions of the Arctic that result in differing descriptions of the land and sea areas encompassed by the term. Policy discussions of the Arctic can employ varying definitions of the region, and readers should bear in mind that the definition used in one discussion may differ from that used in another. This CRS report does not rely on any one definition.

Arctic Circle Definition and Resulting Arctic Countries

The most common and basic definition of the Arctic defines the region as the land and sea area north of the Arctic Circle (a circle of latitude at about 66.34° North). For surface locations within this zone, the sun is generally above the horizon for 24 continuous hours at least once per year (at the summer solstice) and below the horizon for 24 continuous hours at least once per year (at the winter solstice).

The Arctic Circle definition includes the northernmost third or so of Alaska, as well as the Chukchi Sea, which separates that part of Alaska from Russia, and U.S. territorial and Exclusive Economic Zone (EEZ) waters north of Alaska. It does not include the lower two-thirds or so of Alaska or the Bering Sea, which separates that lower part of the state from Russia.

Eight countries have territory north of the Arctic Circle: the United States (Alaska), Canada, Russia, Norway, Denmark (by virtue of Greenland, a member country of the Kingdom of Denmark), Finland, Sweden, and Iceland.[2] These eight countries are often referred to as the Arctic countries, and they are the member states of the Arctic Council, an intergovernmental forum established in 1996.[3] A subset of the eight Arctic countries are the five countries that are considered Arctic coastal states: the United States, Canada, Russia, Norway, and Denmark (by virtue of Greenland).

Definition in Arctic Research and Policy Act (ARPA) of 1984

Section 112 of the Arctic Research and Policy Act (ARPA) of 1984 (Title I of P.L. 98-373 of July 31, 1984)[4] defines the Arctic as follows:

> As used in this title, the term "Arctic" means all United States and foreign territory north of the Arctic Circle and all United States territory north and west of the boundary formed by the Porcupine, Yukon, and Kuskokwim Rivers [in Alaska]; all contiguous seas, including the Arctic Ocean and the Beaufort, Bering, and Chukchi Seas; and the Aleutian chain.

This definition, which is codified at 15 U.S.C. 4111,[5] includes certain parts of Alaska below the Arctic Circle, including the Aleutian Islands and portions of central and western mainland Alaska, such as the Seward Peninsula and the Yukon Delta.

Figure 1 below shows the Arctic area of Alaska as defined by ARPA.

Figure 2 shows the entire Arctic area as defined by ARPA.

Other Definitions

Other definitions of the Arctic are based on factors such as average temperature, the northern tree line, the extent of permafrost on land, the extent of sea ice on the ocean, or jurisdictional or administrative boundaries.[6] A definition based on a climate-related factor could circumscribe differing areas over time as a result of climate change.

The 10° C isotherm definition of the Arctic defines the region as the land and sea area in the northern hemisphere where the average temperature for the warmest month (July) is below 10° Celsius, or 50° Fahrenheit. This definition results in an irregularly shaped Arctic region that excludes some land and sea areas north of the Arctic Circle but includes some land and sea areas south of the Arctic Circle. This definition currently excludes all of Finland and

Sweden, as well as some of Alaska above the Arctic Circle, while including virtually all of the Bering Sea and Alaska's Aleutian Islands.[7]

The definition of the Arctic adopted by the Arctic Monitoring and Assessment Programme (AMAP)—a working group of the Arctic Council—"essentially includes the terrestrial and marine areas north of the Arctic Circle (66°32' N), and north of 62° N in Asia and 60° N in North America, modified to include the marine areas north of the Aleutian chain, Hudson Bay, and parts of the North Atlantic, including the Labrador Sea."[8] The AMAP website includes a map showing the Arctic Circle, 10° C isotherm, tree line, and AMAP definitions of the Arctic.[9]

Some observers use the term "high north" as a way of referring to the Arctic. Some observers make a distinction between the "high Arctic"—meaning, in general, the colder portions of the Arctic that are closer to the North Pole—and other areas of the Arctic that are generally less cold and further away from the North Pole, which are sometimes described as the low Arctic or the subarctic.

Source: U.S. Arctic Research Commission (http://www.arctic.gov/maps/ARPA_Alaska_only_150dpi.jpg, accessed on December 23, 2011).

Figure 1. Arctic Area of Alaska as Defined by ARPA.

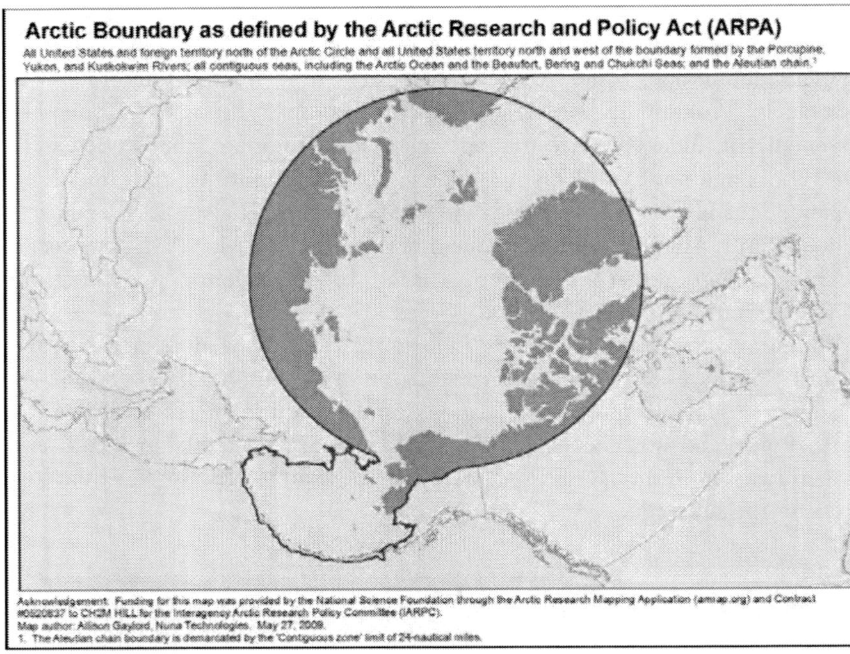

Source: U.S. Arctic Research Commission (http://www.arctic.gov/maps/ARPA_Polar_150dpi.jpg, accessed on December 23, 2011).

Figure 2. Entire Arctic Area as Defined by ARPA.

U.S. Arctic Research

Arctic Research and Policy Act (ARPA) of 1984, As Amended

The Arctic Research and Policy Act (ARPA) of 1984 (Title I of P.L. 98-373 of July 31, 1984)[10] "provide[s] for a comprehensive national policy dealing with national research needs and objectives in the Arctic."[11] The act, among other things

- made a series of findings concerning the importance of the Arctic and Arctic research;
- established the U.S. Arctic Research Commission (USARC) to promote Arctic research and recommend Arctic research policy;
- designated the National Science Foundation (NSF) as the lead federal agency for implementing Arctic research policy;

- established the Interagency Arctic Research Policy Committee (IARPC) to develop a national Arctic research policy and a five-year plan to implement that policy, and designated the NSF representative on the IARPC as its chairperson;[12] and
- defined the term "Arctic" for purposes of the act.

The Arctic Research and Policy Act of 1984 was amended by P.L. 101-609 of November 16, 1990. For the texts of the Arctic Research and Policy Act of 1984 and P.L. 101-609, see **Appendix A** and **Appendix B**, respectively.

FY2014 NSF Budget Request for Arctic Research

NSF—the lead federal agency for implementing Arctic research policy (see "Arctic Research and Policy Act (ARPA) of 1984, As Amended")—carries out Arctic research activities through its Division of Polar Programs (PLR), which is part of its Directorate for Geosciences (GEO).[13] NSF is requesting a total of $464.9 million for PLR for FY2014, including $139.3 million for research in both the Arctic and Antarctic and $44.0 million for Arctic research support and logistics.[14] NSF states in its FY2014 overview of GEO that

> As the primary U.S. supporter of fundamental research in the polar regions, GEO provides interagency leadership for U.S. polar activities. In the Arctic, NSF helps coordinate research planning as directed by the Arctic Research Policy Act of 1984. The NSF Director chairs the Interagency Arctic Research Policy Committee created for this purpose, which is now a component of the President's National Science and Technology Council.[15]

NSF states in its FY2014 overview of PLR that

> The Division of Polar Programs (PLR) is the primary U.S. supporter of, and serves NSF interagency leadership responsibilities for, fundamental research in the polar regions. The Arctic Sciences section supports research in social, earth systems, and a broad range of natural sciences; its Research Support & Logistics program is driven by and responds to research by assisting researchers with access to the Arctic and for engagement in planning and sharing of results with local Arctic communities....
>
> PLR's FY 2014 Request reflects three key priorities: (1) maintaining strong disciplinary programs that provide a base for investments in cross-

> disciplinary science programs; (2) focusing basic research on cross-foundation (e.g., Science, Engineering, and Education for Sustainability (SEES), Cyberinfrastructure Framework for 21st Century Science, Engineering, and Education (CIF21), GEOTRACES), and interagency priorities; and (3) supporting and improving the efficiency of critical facilities that enable research in both polar regions.... For the Arctic, shared cross-directorate basic research objectives, the recently released Interagency Arctic Research Policy Committee's (IARPC) Arctic Research Plan FY 2013-2017 (Five-Year Plan) and the National Ocean Policy (NOP) Implementation Strategy inform science investment priorities.
>
> In general, 40 percent of the PLR research portfolio is available for new grants and 60 percent for continuing grants. Approximately 70 percent of the overall budget supports facilities and logistics.[16]

NSF states that, compared to the FY2012-enacted funding level, changes reflected in the FY2014 request for $139.4 million for Arctic and Antarctic research activities within PLR include, among other things,

- Funding increases to develop mechanisms to coordinate with Alaska Native organizations to build partnerships and broaden participation in scientific research, and to strengthen PLR's long-standing commitment to engaging Alaska Native students and communities. (+$200,000, to a total of $300,000)
- PLR will collaborate with OCE [the Division of Ocean Sciences within GEO] to support the U.S. components of several marine expeditions under the GEOTRACES program to measure trace elements and isotopes across the Arctic Ocean and the connecting passages between the Arctic, Atlantic, and Pacific Oceans. (+$2.0 million, to a total of $2.0 million)
- Core Arctic research will be prioritized to support reinvestment in sustained synthesis of modeling and observational data at the system level and for research on the Chukchi and Beaufort Seas, also taking an integrative system-level approach. These will complement the more focused research supported by mission agencies, such as the Bureau of Ocean Energy Management (BOEM) and the National Oceanic and Atmospheric Administration (NOAA), and contribute to the coordinated activity under IARPC. ($6.0 million)[17]

Regarding its FY2014 request for $44.0 million for Arctic research support and logistics within PLR, NSF states:

> The Arctic Research Support and Logistics program provides support for Arctic researchers, including airplanes, helicopters, access to icebreakers and field camps for researchers participating in approximately 150 projects in remote sites in Alaska, Canada, Arctic Scandinavia, Russia, and the Arctic Ocean. Summit Station on the Greenland icecap operates as a year-round international site for a variety of atmospheric and geophysical measurements. Increased funding [requested for FY2014, compared to the enacted FY2012 level] will support anticipated ship-time on the new UNOLS [University National Oceanographic Laboratory System] vessel SIKULIAQ. (+$460,000, to a total of $460,000).[18]

January 2009 Arctic Policy Directive (NSPD 66/HSPD 25)

On January 12, 2009, the George W. Bush Administration released a presidential directive establishing a new U.S. policy for the Arctic region. The directive, dated January 9, 2009, was issued as National Security Presidential Directive 66/Homeland Security Presidential Directive 25 (NSPD 66/HSPD 25). The directive was the result of an interagency review, and it superseded for the Arctic (but not the Antarctic) a 1994 presidential directive on Arctic and Antarctic policy. The directive, among other things,

- states that the United States is an Arctic nation, with varied and compelling interests in the region;
- sets forth a six-element overall U.S. policy for the region;
- describes U.S. national security and homeland security interests in the Arctic; and
- discusses a number of issues as they relate to the Arctic, including international governance, the extended continental shelf and boundary issues; promotion of international scientific cooperation; maritime transportation; economic issues, including energy; and environmental protection and conservation of natural resources.

The Obama Administration has not issued a new directive superseding NSPD 66/HSPD 25; it is currently operating under the Bush Administration's policy directive.[19] For the text of NSPD 66/HSPD 25, see **Appendix C**.

May 2010 National Security Strategy

In May 2010, the Obama Administration released a national security strategy document that states:

> The United States is an Arctic Nation with broad and fundamental interests in the Arctic region, where we seek to meet our national security needs, protect the environment, responsibly manage resources, account for indigenous communities, support scientific research, and strengthen international cooperation on a wide range of issues.[20]

May 2013 National Strategy for Arctic Region

On May 10, 2013, the Obama Administration released a document entitled *National Strategy for the Arctic Region.*[21] The document appears to supplement rather than supersede the January 2009 Arctic policy directive (NSPD 66/HSPD 25) discussed above.[22]

The executive summary of *National Strategy for the Arctic Region* begins by quoting the above statement from the May 2010 national security strategy document, and then states:

> The *National Strategy for the Arctic Region* sets forth the United States Government's strategic priorities for the Arctic region. This strategy is intended to position the United States to respond effectively to challenges and emerging opportunities arising from significant increases in Arctic activity due to the diminishment of sea ice and the emergence of a new Arctic environment. It defines U.S. national security interests in the Arctic region and identifies prioritized lines of effort, building upon existing initiatives by Federal, state, local, and tribal authorities, the private sector, and international partners, and aims to focus efforts where opportunities exist and action is needed. It is designed to meet the reality of a changing Arctic environment, while we simultaneously pursue our global objective of combating the climatic changes that are driving these environmental conditions. Our strategy is built on three lines of effort:

1. **Advance United States Security Interests** – We will enable our vessels and aircraft to operate, consistent with international law, through, under, and over the airspace and waters of the Arctic, support lawful commerce, achieve a greater awareness of activity in the region, and intelligently evolve our Arctic infrastructure and capabilities, including ice-capable platforms as needed. U.S. security

in the Arctic encompasses a broad spectrum of activities, ranging from those supporting safe commercial and scientific operations to national defense.

2. **Pursue Responsible Arctic Region Stewardship** – We will continue to protect the Arctic environment and conserve its resources; establish and institutionalize an integrated Arctic management framework; chart the Arctic region; and employ scientific research and traditional knowledge to increase understanding of the Arctic.
3. **Strengthen International Cooperation** – Working through bilateral relationships and multilateral bodies, including the Arctic Council, we will pursue arrangements that advance collective interests, promote shared Arctic state prosperity, protect the Arctic environment, and enhance regional security, and we will work toward U.S. accession to the United Nations Convention on the Law of the Sea (Law of the Sea Convention).

Our approach will be informed by the following guiding principles:

- **Safeguard Peace and Stability** – Seek to maintain and preserve the Arctic region as an area free of conflict, acting in concert with allies, partners, and other interested parties. Support and preserve: international legal principles of freedom of navigation and overflight and other uses of the sea and airspace related to these freedoms, unimpeded lawful commerce, and the peaceful resolution of disputes for all nations.
- **Make Decisions Using the Best Available Information** – Across all lines of effort, decisions need to be based on the most current science and traditional knowledge.[23]
- **Pursue Innovative Arrangements** – Foster partnerships with the state of Alaska, Arctic states, other international partners, and the private sector to more efficiently develop, resource, and manage capabilities, where appropriate and feasible, to better advance our strategic priorities in this austere fiscal environment.
- **Consult and Coordinate with Alaska Natives** – Engage in a consultation process with Alaska Natives, recognizing tribal governments' unique legal relationship with the United States and providing for meaningful and timely opportunity to inform Federal policy affecting Alaskan Native communities.[24]

The Arctic and the U.N. Convention on Law of the Sea (UNCLOS)[25]

Background to UNCLOS

In November 1994, the United Nations Convention on the Law of the Sea (UNCLOS) entered into force. This convention establishes a treaty regime to govern activities on, over, and under the world's oceans. It builds on the four 1958 law of the sea conventions and sets forth a framework for future activities in parts of the oceans that are beyond national jurisdiction.[26] The 1982 Convention and its 1994 Agreement relating to Implementation of Part XI of the Convention were transmitted to the Senate on October 6, 1994.[27] In the absence of Senate advice and consent to adherence, the United States is not a party to the convention and agreement.

Part VI of UNCLOS and Commission on Limits of Continental Shelf

Part VI of the convention, dealing with the Continental Shelf, and Annex II, which established a Commission on the Limits of the Continental Shelf, are most pertinent to the Arctic as it becomes more accessible ocean space, bordered by five coastal states.[28] The convention gives the coastal state sovereign jurisdiction over the resources, including oil and gas, of its continental shelf.[29] Under Article 76 of the convention, a coastal state with a broad continental margin may establish a shelf limit beyond 200 nautical miles. This jurisdiction is subject to the submission of the particulars of the intended limit and supporting scientific and technical data by the coastal state to the commission for review and recommendation.[30] The commission reviews the documentation and, by a two-thirds majority, approves its recommendations to the submitting state. Coastal states agree to establish the outer limits of their continental shelf, in accordance with this process and with their national laws. In instances of disagreement with the commission's recommendations, the coastal state may make a revised or new submission. The actions of the commission "shall not prejudice matters relating to delimitation of boundaries between States with opposite or adjacent coasts."[31] The "limits established by a coastal State on the basis of these recommendations shall be final and binding."[32]

U.S. Activities as a Non-Party to UNCLOS

As a non-party to the convention, the United States cannot participate as a member of the commission; it cannot submit a claim under Article 76. Over the years, however, it has submitted observations on submissions made by other states, requesting that those observations be made available online and to the commission. In addition, since 2001, the United States has gathered and analyzed data to determine the outer limits of its extended continental shelf. Starting in 2007, this effort became the Extended Continental Shelf Project, directed by an interagency task force under the lead of the Department of State.[33]

Some observers have suggested that a separate regime be negotiated to address the changing circumstances in the Arctic. They maintain that this phenomenon was not envisioned at the time the Law of the Sea Convention was negotiated. Still others suggest that the Arctic region above a certain parallel be designated a wilderness area, and they cite as precedent Article 4 of The Antarctic Treaty, under which any current claims to sovereign territory are frozen and

> No acts or activities taking place while the present Treaty is in force shall constitute a basis for asserting, supporting or denying a claim to territorial sovereignty in Antarctica or create any rights of sovereignty in Antarctica. No new claim, or enlargement of an existing claim, to territorial sovereignty in Antarctica shall be asserted while the present Treaty is in force.

Supporters of the Law of the Sea Convention maintain that changing circumstances in the Arctic strengthen their argument that the United States should become a party to the convention. In this way, they argue, the United States can be best situated to protect and serve its national interests, under both Article 76 and other parts of the convention.

ISSUES FOR CONGRESS

Climate Change and Loss of Arctic Sea Ice[34]

Record low extents of Arctic sea ice in 2012 and 2007 have focused scientific and policy attention on climate changes in the High North, and to the implications of projected ice-free[35] seasons in the Arctic within decades. The Arctic has been projected by several scientists to be ice-free in most late summers as soon as the 2030s.[36] This opens opportunities for transport through the Northwest Passage and the Northern Sea Route, extraction of potential oil and gas resources, and expanded fishing and tourism (**Figure 3**).

More broadly, physical changes in the Arctic include warming ocean, soil, and air temperatures; melting permafrost; shifting vegetation and animal abundances; and altered characteristics of Arctic cyclones. All these changes are expected to affect traditional livelihoods and cultures in the region and survival of polar bear and other animal populations, and raise risks of pollution, food supply, safety, cultural losses, and national security. Moreover, linkages ("teleconnections") between warming Arctic conditions and extreme events in the mid-latitude continents are increasingly evident, identified in such extreme events as the heat waves and fires in Russia in 2010; severe winters in the eastern United States and Europe in 2009/2010 and in Europe in 2011/2012;[37] and Indian summer monsoons and droughts. Hence, changing climate in the Arctic suggests important implications both locally and across the Hemisphere.

Like the rest of the globe, temperatures in the Arctic have varied[38] but show a significant warming trend since the 1970s, and particularly since 1995.[39] The annual average temperature for the Arctic region (from 60° to 90° N) is now about 1.8° F warmer than the "climate normal" (the average from 1961 to 1990). Temperatures in October-November are now about 9° F above the seasonal normal. Scientists have concluded that most of the *global* warming of the last three decades is very likely caused by human-related emissions of greenhouse gases (GHG, mostly carbon dioxide); they expect the GHG-induced warming to continue for decades, even if, and after, GHG concentrations in the atmosphere have been stabilized. The extra heat in the Arctic is amplified by processes there (the "polar amplification") and may result in irreversible changes on human timescales.

The observed warmer temperatures along with rising cyclone size and strength in the Arctic have reduced sea ice extent, thickness, and ice that persists year-round ("perennial ice"); natural climate variability has likely contributed to the record low ice extents of 2007 and 2012. The 2007 minimum sea ice extent was influenced by warm Arctic temperatures and warm, moist winds blowing from the North Pacific into the central Arctic, contributing to melting and pushing ice toward and into the Atlantic past Greenland. Warm winds did not account for the near-record sea ice minimum in 2008.[40] In early August 2012, an unusually large storm with low pressure developed over the Arctic, helping to disperse the already weak ice into warmer waters and accelerating its melt rate. By August 24, 2012, sea ice extent had shrunk below the previous observed minimum of late September 2007.[41] Modeling of GHG-induced climate change is particularly challenging for the Arctic, but it consistently projects warming through the 21st century, with annual average Arctic temperature increases ranging from +1° to +9.0° C (+2° to +19.0° F), depending on the GHG scenario and model used. While such warming is projected by most models throughout the Arctic, some models project slight cooling localized in the North Atlantic Ocean just south of Greenland and Iceland. Most warming would occur in autumn and winter, "with very little temperature change projected over the Arctic Ocean" in summer months.[42] Due to observed and projected climate change, scientists have concluded that the Arctic will have changed from an ice-covered environment to a recurrent ice-free[43] ocean (in summers) as soon as the late 2030s. The character of ice cover is expected to change as well, with the ice being thinner, more fragile, and more regionally variable. The variability in recent years of both ice quantity and location could be expected to continue.

Territorial Claims and Sovereignty Issues[44]

Motivated in part by a desire to exercise sovereign control over the Arctic region's increasingly accessible oil and gas reserves (see "Oil, Gas, and Mineral Exploration"), the four Arctic coastal states other than the United States—Canada, Russia, Norway, and Denmark (of which Greenland is a territory)—are in the process of preparing territorial claims in the Arctic, including claims for expanded Exclusive Economic Zones (EEZs), for submission to the Commission on the Limits of the Continental Shelf. (As mentioned earlier—see "U.S. Activities As a Non-Party to UNCLOS"—the United States, as a non-party to UNCLOS, cannot participate as a member of the commission; it cannot submit a claim under Article 76. Over the years, however, it has submitted observations on submissions made by other states, requesting that those observations be made available online and to the commission. In addition, since 2001, the United States has gathered and analyzed data to determine the outer limits of its extended continental shelf.)

Russia has been attempting to chart the Arctic Ocean's enormous underwater Lomonosov Ridge in an attempt to show that it is an extension of Russia's continental margin. The Russian claim to this ridge, if accepted, would reportedly grant Russia nearly one-half of the Arctic area; a 2001 claim submitted by Russia was rejected as insufficiently documented. Canada also claims a portion of the Lomonosov Ridge as part of its own underwater continental shelf.[45] In August 2007, a Russian submersible on a research expedition deposited an encased Russian Federation flag on the seabed of the presumed site of the North Pole. The action captured worldwide attention, but analysts note that it did not constitute an official claim to the territory and was therefore a purely symbolic act.

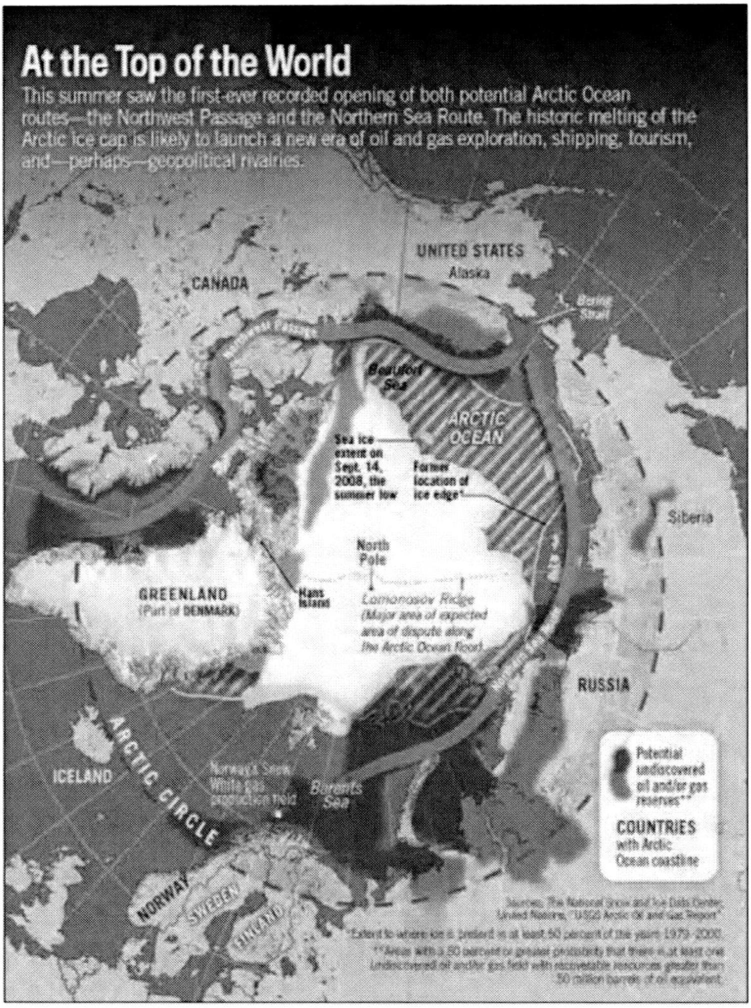

Source: Graphic by Stephen Rountree at *U.S. News and World Report*, http://www.usnews.com/articles/news/world/2008/10/09/global-warming-triggers-an-international-race-for-the-artic/photos/#1.

Figure 3. Arctic Sea Ice Extent in September 2008, Compared with Prospective Shipping Routes and Oil and Gas Resources.

At a May 2008 meeting in Ilulissat, Greenland, the five Arctic coastal states reaffirmed their commitment to the UNCLOS legal framework for the establishment of extended continental shelf limits in the Arctic.[46] (For further discussion, see "Extent of the Continental Margin" in "Oil, Gas, and Mineral Exploration.")

In addition to this process, there are four unresolved Arctic territorial disputes:

- Scientists have forecast that in coming decades, global warming will reduce the ice pack in Canada's northern archipelago sufficiently to permit ships to use the trans-Arctic shipping route known as the Northwest Passage during the summer months (see "Commercial Sea Transportation"). The prospect of such traffic raises a major jurisdictional question. Ottawa maintains that such a passage would be an inland waterway, and would therefore be sovereign Canadian territory subject to Ottawa's surveillance, regulation, and control. The United States, the European Union, and others assert that the passage would constitute an international strait between two high seas.
- The United States and Canada are negotiating over a binational boundary in the Beaufort Sea.
- The United States and Russia in 1990 signed an agreement regarding a disputed area of the Bering Sea; the U.S. Senate ratified the pact the following year, but the Russian *Duma* has yet to approve the accord.
- Denmark and Canada disagree over which country has the territorial right to Hans Island, a tiny, barren piece of rock between Greenland and Canada's Ellesmere Island. Some analysts believe the two countries are vying for control over a future sea lane that might be created if the Arctic ice were to melt sufficiently to create a Northwest Passage. Others claim that the governments are staking out territorial claims in the event that future natural resource discoveries make the region economically valuable.[47]

In addition to these disputes, Norway and Russia had been at odds for decades over the boundary between the two in the so-called "Grey Zone" in the Barents Sea, an area believed to hold rich undersea deposits of petroleum. On September 15, 2010, Norwegian Prime Minister Jens Stoltenberg and Russian President Dmitry Medvedev signed an agreement in Murmansk, a Russian city near the Norwegian border. The accord awards roughly half of the 175,000-square- kilometer area to each country; it spells out fishing rights, and provides for the joint development of future oil and gas finds that straddle the boundary line. Some observers believe it is noteworthy that Russia would concede sovereignty over such a large, resource-rich territory to a small, neighboring country. But others have noted that Moscow may be hoping for Norwegian cooperation in developing offshore resources, and eventually in winning approval when Russia submits its Article 76 UNCLOS claim.[48]

In August 2010, Canadian Foreign Minister Lawrence Cannon announced a new "Statement of Canada's Arctic Policy," which reaffirmed the government's commitment to Canada's sovereignty in the region, to economic and social development, to environmental protection, and to empowerment of the peoples in the north. The statement also emphasized the government's intention to negotiate settlements to its disputes with the United States over the Beaufort Sea boundary, and with Denmark over Hans Island. Minister Cannon declared that "making progress on outstanding boundary issues will be a top priority."[49] Also, despite their dispute over Hans Island, Canada and Denmark have been working together on Arctic issues. In May 2010, the two countries' military chiefs of staffs signed a memorandum of understanding on Arctic Defense, Security, and Operational Cooperation, committing the two countries to "enhanced consultation, information exchange, visits, and exercises."[50]

Commercial Sea Transportation[51]

Background

The search for a shorter route from the Atlantic to Asia has been the quest of maritime powers since the Middle Ages. The melting of Arctic ice raises the possibility of saving several thousands of miles and several days of sailing between major trading blocs.[52] If the Arctic were to become a viable shipping route, the ramifications could extend far beyond the Arctic. For example, lower shipping costs could be advantageous for China (at least its northeast region), Japan, and South Korea because their manufactured products exported to Europe or North America could become less expensive relative to other emerging manufacturing centers in Southeast Asia, such as India.[53] Melting ice could potentially open up two trans-Arctic routes (see **Figure 3**):[54]

- **The Northern Sea Route** (NSR, a.k.a. the "Northeast Passage"), along Russia's northern border from Murmansk to Provideniya, is about 2,600 nautical miles in length. It was opened by the Soviet Union to domestic shipping in 1931 and to transit by foreign vessels in 1991. Even so, these waters are little used by non- Russian ships.[55] This route would be applicable for trade between northeast Asia (north of Singapore) and northern Europe.

- **The Northwest Passage** (NWP) runs through the Canadian Arctic Islands. The NWP actually consists of several potential routes. The southern route is through Peel Sound in Nunavut, which has been open in recent summers and contains mostly one-year ice. However, this route is circuitous, contains some narrow channels, and is shallow enough to impose draft restrictions on ships. The more northern route, through McClure Strait from Baffin Bay to the Beaufort Sea north of Alaska, is much more direct, and therefore more appealing to ocean carriers, but more prone to ice blockage.[56] The NWP is potentially applicable for trade between northeast Asia (north of Shanghai) and the northeast of North America, but may be less commercially viable than the NSR.[57]

Destination Traffic, Not Trans-Arctic Traffic

Almost all cargo ship activity currently taking place in the Arctic is to transport natural resources from the Arctic or to deliver general cargo and supplies to communities and natural resource extraction facilities. Thus, cargo ship traffic in the Arctic presently is mostly regional, not trans- Arctic. While there has been a recent uptick in Arctic shipping activity, this activity has more to do with a spike in commodity prices than it does with the melting of Arctic ice. Even so, recent activity is less than it has been in the past. The NSR continues to account for the bulk of Arctic shipping activity. The western end of the NSR, in the Barents Sea, is open year round. Further east, sections of the NSR are open for as little as two and a half months during the summer. However, in 2011, the route was open for five months and 34 ships, accompanied by a Russian icebreaker, passed through.[58]

Cruise Ship Activity

Considerable cruise ship activity takes place in Arctic waters. In the summer of 2007, three cruise ships reportedly sailed through the NWP from the Atlantic to Alaska's North Slope.[59] In August 2010, a cruise ship with over 150 passengers ran aground in the NWP. In the Barents Sea, there are regular cruise ships sailing to Svalbard. The inherent dangers for passenger ships in the Arctic have prompted calls for international regulations promoting the safety of cruise ships in the area. Some have suggested that cruise ships sail in pairs to provide assistance to one another, given the Arctic's remoteness and the difficulty land-based rescuers would have in reaching a vessel in distress.[60] Requiring that Arctic cruise vessels have ice-strengthened hulls and be equipped with enclosed lifeboats could be other safety requirements. In 2003, some Arctic cruise and tourist operators formed the Association of Arctic Expedition Cruise Operators (AECO) to establish agreed-upon safety and environmental protection guidelines, but this organization only covers the portion of the Arctic around Greenland, Svalbard, and Jan Mayen.[61]

Unpredictable Ice Conditions Hinder Trans-Arctic Shipping

Arctic waters do not necessarily have to be ice free to be open to shipping. Multiyear ice can be over 10 feet thick and problematic even for icebreakers, but one-year ice is typically 3 feet thick or less. This thinner ice can be more readily broken up by icebreakers or ice class ships (cargo ships with reinforced hulls and other features for navigating in ice-infested waters). However, more open water in the Arctic has resulted in another potential obstacle to shipping: unpredictable ice flows. In the NWP, melting ice and the opening of waters that were once covered with one- year ice has allowed blocks of multiyear ice from farther north, or icebergs from Greenland, to flow into potential sea lanes. The source of this multiyear ice is not predicted to dissipate in spite of climate change. Moreover, the flow patterns of these ice blocks are very difficult to predict, and they have floated into potential routes for shipping.[62] Thus, the lack of ice in potential sea lanes during the summer months can add even greater unpredictability to Arctic shipping. This is in addition to the extent of ice versus open water, which is also highly variable from one year to the next and seasonally.

The unpredictability of ice conditions is a major hindrance for trans-Arctic shipping in general, but can be more of a concern for some types of ships than it is for others. For instance, it would be less of a concern for cruise ships, which may have the objective of merely visiting the Arctic rather than passing through and could change their route and itinerary depending on ice conditions. On the other hand, unpredictability is of the utmost concern for container ships that carry thousands of containers from hundreds of different customers, all of whom expect to unload or load their cargo upon the ship's arrival at various ports as indicated on the ship's advertised schedule. The presence of even small blocks of ice or icebergs from a melting Greenland ice sheet requires slow sailing and could play havoc with schedules. Ships carrying a single commodity in bulk from one port to another for just one customer have more flexibility in terms of delivery windows, but would not likely risk an Arctic passage under prevailing conditions.

Ice is not the sole impediment to Arctic shipping. The region frequently experiences adverse weather, including not only severe storms, but also intense cold, which can impair deck machinery. During the summer months when sea lanes are open, heavy fog is common in the Arctic.

Commercial ships would face higher operating costs on Arctic routes than elsewhere. Ship size is an important factor in reducing freight costs. Many ships currently used in other waters would require two icebreakers to break a path wide enough for them to sail through; ship owners could reduce that cost by using smaller vessels in the Arctic, but this would raise the cost per container or per ton of freight.[63] Also, icebreakers or ice-class cargo vessels burn more fuel than ships designed for more temperate waters and would have to sail at slower speeds. The shipping season in the Arctic only lasts for a few weeks, so icebreakers and other special required equipment would sit idle the remainder of the year. None of these impediments by themselves may be enough to discourage Arctic passage but they do raise costs, perhaps enough to negate the savings of a shorter route. Thus, from the perspective of a shipper or a ship owner, shorter via the Arctic does not necessarily mean cheaper and faster.[64]

Basic Navigation Infrastructure Is Lacking

Considerable investment in navigation-related infrastructure would be required if trans-Arctic shipping were to become a reality. Channel marking buoys and other floating visual aids are not possible in Arctic waters because moving ice sheets will continuously shift their positions. Therefore, vessel captains would need to rely on marine surveys and ice charts. For some areas in the Arctic, however, these surveys and charts are out of date and not sufficiently accurate.[65] To remedy this problem, aviation reconnaissance of ice conditions and satellite images would need to become readily available for ship operators.[66] Ship-to-shore communication infrastructure would need to be installed where possible. Refueling stations may be needed, as well as, perhaps, transshipment ports where cargo could be transferred to and from ice-capable vessels at both ends of Arctic routes. Shipping lines would need to develop a larger pool of mariners with ice navigation experience. Marine insurers would need to calculate the proper level of risk premium for polar routes, which would require more detailed information about Arctic accidents and incidents in the past.

The U.S. Army Corps of Engineers, along with the state of Alaska, is studying the feasibility of a "deep-draft" port in the Arctic (accommodating ships with a draft of up to 35 feet). The northern and northwestern coastlines of Alaska are exceptionally shallow, generally limiting harbor and near shore traffic to shallow-draft barges. Coast Guard cutters and icebreakers have drafts of 35 to 40 feet while NOAA research vessels have drafts of 16 to 28 feet, so at present these vessels are based outside the Arctic and must sail considerable distances to reach Arctic duty stations. Supply vessels supporting offshore oil rigs typically have drafts over 20 feet. A deep-draft port could serve as a base of operations for larger vessels, facilitating commercial maritime traffic in the Arctic.[67]

Another study on maritime infrastructure needs is being conducted by the U.S. Committee on the Marine Transportation System, a Cabinet-level committee of federal agencies with responsibilities for marine transportation. It is due to report to Congress in 2012 a priority of federal actions the committee believes are needed to ensure maritime safety, environmental protection, and national security in the Arctic.[68]

Regulation of Arctic Shipping

Due to the international nature of the shipping industry, maritime trading nations have adopted international treaties that establish standards for ocean carriers in terms of safety, pollution prevention, and security. These standards are agreed upon by shipping nations through the International Maritime Organization (IMO), a United Nations agency that first met in 1959.[69]

Key conventions that the 168 IMO member nations have adopted include the Safety of Life at Sea Convention (SOLAS), which was originally adopted in response to the *Titanic* disaster in 1912 but has since been revised several times; the Prevention of Pollution from Ships (MARPOL), which was adopted in 1973 and modified in 1978; and the Standards for Training, Certification, and Watchkeeping for Seafarers (SCTW), which was adopted in 1978 and amended in 1995. It is up to ratifying nations to enforce these standards. The United States is a party to these conventions, and the U.S. Coast Guard enforces them when it boards and inspects ships and crews arriving at U.S. ports and the very few ships engaged in international trade that sail under the U.S. flag.

Like the United States, most of the other major maritime trading nations lack the ability to enforce these regulations as a "flag state" because much of the world's merchant fleet is registered under so-called "flags of convenience." While most ship owners and operators are headquartered in developed countries, they often register their ships in Panama, Liberia, the Bahamas, the Marshall Islands, Malta, and Cyprus, among other "open registries," because these nations offer more attractive tax and employment regulatory regimes. Because of this development, most maritime trading nations enforce shipping regulations under a "port state control" regime—that is, they require compliance with these regulations as a condition of calling at their ports. The fragmented nature of ship ownership and operation can be a further hurdle to regulatory enforcement. It is common for cargo ships to be owned by one company, operated by a second company (which markets the ship's space), and managed by a third (which may supply the crew and other services a ship requires to sail), each of which could be headquartered in different countries.

Arctic Guidelines

While SOLAS and other IMO conventions include provisions regarding the operation of ships in ice-infested waters, they are not specific to the polar regions. To supplement existing requirements, in December 2002, the IMO approved guidelines for ships operating in Arctic ice- covered waters.[70] These were only recommendations for ships operating in the Arctic, not requirements. They apply to passenger and cargo ships of 500 gross tons or more engaged in international voyages. They do not apply to fishing vessels, military vessels, pleasure yachts, and smaller cargo ships. The guidelines are intended to improve safety and prevent pollution in the Arctic, and they include provisions on ship construction, ship equipment related to navigation, and crew training and ship operation. The guidelines recommend that ships carry fully enclosed lifeboats or carry tarpaulins to cover their lifeboats. They recommend that each crew include at least one ice navigator with documented evidence of having completed an ice navigation training program. The IMO is in the process of drafting mandatory requirements for ships operating in the Arctic.[71]

Nations can enforce additional requirements on ships arriving at their ports or sailing through their coastal waters. For instance, U.S. Coast Guard regulations largely follow IMO conventions but mandate additional requirements in some areas. U.S. coastal states can require ships calling at their ports to take additional safety and pollution prevention safeguards. Canada and Russia have additional pollution regulations for Arctic waters exceeding MARPOL. The U.S. Coast Guard is seeking agreement with Russia to establish a vessel traffic separation scheme for the Bering Strait between Alaska and Russia, which now experiences over 300 transits per year.[72]

Oil, Gas, and Mineral Exploration[73]

Decreases in summer polar ice will likely allow for more options for exploring for oil and gas in certain offshore areas. Similarly, shrinking glaciers onshore could expose land containing economic deposits of gold, iron ore, or other minerals that were previously covered by glacial ice. Currently offshore areas encompassing more than 1 billion acres and more than 6,000 miles of coastline—more coastline than in the rest of the United States combined—are considered potential areas for offshore energy development. These Arctic regions include the Beaufort and Chukchi Seas, the Bering Sea, Cook Inlet, and the Gulf of Alaska.

Warming that causes permafrost to melt could pose challenges to onshore exploration activities because ground structures, such as pipelines and other infrastructure that depend on footings sunk into the permafrost for support, could become unstable. Despite the warming temperatures, however, exploration and development in the Arctic would still be subject to harsh conditions, especially in winter, which makes it costly and challenging to develop infrastructure necessary to produce, store, and transport oil, gas, and minerals from newly discovered deposits. In offshore areas, severe weather poses challenges to the ongoing operation of federally regulated operations involving six producing wells.[74]

Shrinking sea ice cover in the Arctic has also intensified interest in surveying and mapping the continental margins of countries with lands in the Arctic. Delineating the extent of the continental margins beyond the 200 nautical mile Exclusive Economic Zone (EEZ) could lead to considering development on substantial amounts of submerged lands.[75] Mapping projects are underway, by individual countries and through cooperative government studies, to support national claims to submerged lands which may contain large amounts of oil, natural gas, methane hydrates, or minerals.

One historical determinant of energy resource management in the Arctic has been federal moratoriums enacted by Congress on an annual basis (as part of DOI appropriations acts) and issued by the President through treaties or executive orders.[76] While not affecting all Alaska operations, these moratoriums effectively banned federally regulated planning and permitting in the Bristol Bay area of the North Aleutian Basin. Since Congress allowed most statutory bans in the region to expire in 2004,[77] federal policy on development rights and permits governing access to the Arctic for drilling has been in flux. In 2010, President Obama reinstated a moratorium in the North Aleutian Basin, withdrawing acreage located in Bristol Bay from eligibility for oil and gas leasing until after 2017.[78] Pursuant to this executive ban, gathering geological and geophysical data and other development-related activities are prohibited in Bristol Bay.

Federal officials report regional operations related to the six producing wells are stable[79] and that plans are unfolding for oil and gas leasing activity in the future.[80] Prompted by claims voiced by some Alaska state officials and energy, trade, and shipping groups favoring greater geographical diversity for federally regulated offshore energy projects, President Obama and many in Congress have expressed interest in expanding America's ocean energy portfolio in the region.[81] As part of the Administration's strategy to enhance domestic energy production, DOI approved a program referred to as the Five Year Program announcing regulatory actions in offshore areas from 2012 through 2017.[82] This program was formulated in part by lessons learned in the aftermath of the *Deepwater Horizon* events in the Gulf of Mexico.[83] While the program defers activities in the Bristol Bay (pursuant to the executive moratorium), it proposes three lease sales elsewhere in the region: in the Chukchi Sea (2016), Cook Inlet (2016), and the Beaufort Sea (2017).

One example of current offshore permitting activity in the region includes a subsidiary of Shell Oil Company[84] gaining permits for exploratory oil and natural gas drilling operations in the Chukchi and Beaufort Seas during the 2012 drilling season.[85] Public attention to these Shell permits peaked during reports of a December 31, 2012, incident involving a drillship named the *Kulluk*, owned by Shell. Media reports that the drillship ran aground on the shoreline of Sitkalidak Island, AK, sparked attention to safety issues.[86] As part of a response to the grounding, federal safety officials commenced an investigation of the incident.[87] Industry commentators, including Randall Luthi, President of the National Ocean Industries Association (NOIA), noted that drilling in U.S. waters can be achieved safely and that this incident should not be a stumbling block to further development in the region.[88] Other commentators countered that developing Arctic resources is risky and investigating this incident offers guidance for future permitting in the region.[89]

The future of offshore drilling in the region is somewhat dependent on the outcome of judicial proceedings that govern aspects of Shell's permitting process.[90] The future timetable for Shell's permits might also depend on DOI regulatory actions focusing on, among other issues, the *Arctic Challenger* (an Arctic Containment System, ACS), a key component of Shell's drilling plan.[91] As part of listing various operational and management issues in the investigatory report described above, DOI announced further studies are warranted prior to approving future Shell drilling operations. Specifically, in terms of fabricating and testing certain safety systems and establishing the scope of its operational plans, DOI investigators found Shell lacked adequate preparation and Shell's timelines did not allow for full certification of its ACS.[92] Based on the information available to date from the courts and from DOI on the status of the Shell permitting process, it remains to be seen how permitting for drilling in the Arctic might proceed, if at all, with respect to the 2013 drilling season.

Oil and Gas

A primary driver for the increased interest in exploring for oil and gas offshore in the Arctic is the shrinking Arctic ice cap, or conversely, the growing amount of ice-free ocean in the summertime. Reduced sea ice in the summer means that ships towing seismic arrays[93] can explore previously inaccessible regions of the Arctic Ocean, Chukchi Sea, Beaufort Sea, and other offshore regions for longer periods of time without as much risk of colliding with floating sea ice. Less sea ice over longer periods compared to previous decades also means that the seasonal window for offshore drilling in the Arctic remains open longer in the summer and increases the chances for making a discovery.

In addition to the improved access to larger portions of the Arctic afforded by shrinking sea ice, interest in Arctic oil and gas was fueled by a 2008 U.S. Geological Survey (USGS) appraisal of undiscovered oil and gas north of the Arctic Circle.[94] The USGS asserted that the "extensive Arctic continental shelves may constitute the geographically largest unexplored prospective area for petroleum remaining on Earth."[95] In the report, the USGS estimates that 90 billion barrels of oil, nearly 1,700 trillion cubic feet of natural gas, and 44 billion barrels of natural gas liquids may remain to be discovered in the Arctic. An article published in Science magazine indicated that 30% of the world's undiscovered natural gas and 13% of the world's undiscovered oil may be found north of the Arctic Circle.[96]

Despite the warming trend in the Arctic, a discovery of new oil and gas deposits far from existing storage, pipelines, and shipping facilities cannot be developed until infrastructure is built to extract and transport the petroleum. Other areas with oil and gas potential, such as offshore east Greenland, may take even longer to develop because they are still covered by substantial summer sea ice that would hinder exploration and development unless more ice melts.[97] Further, in 2010 Secretary of the Interior Salazar asked the USGS to assess the science needs that would inform the Obama Administration's consideration of how to develop oil and gas resources in the Arctic.[98] The

USGS released a report in response that identified some overarching themes, which include the need for (1) large-scale synthesis of data and information, (2) enhanced dialog and collaborative science planning, and (3) a more transparent and inclusive planning and decision- making process.[99]

Extent of the Continental Margin

Increased interest in developing offshore resources in the Arctic has sparked efforts by nations bordering the Arctic Ocean to map the extent of their continental margins beyond the 200-mile EEZ limit. Under UNCLOS, nations can submit a claim to submerged lands if they demonstrate that their continental margin extends beyond the 200-mile limit according to the definition in Article 76. Under Article 76, the extent of the continental margin beyond the 200-mile limit depends on the position of the foot of the continental slope, the thickness of sediments, and the depth of water. Also, the continental margin could include geologic features that extend from the continent out to sea, which may include undersea ridges continuing for hundreds of miles offshore.

As mentioned earlier—see "U.S. Activities As a Non-Party to UNCLOS"—the United States, as a non-party to UNCLOS, cannot participate as a member of the Commission on the Limits of the Continental Shelf; it cannot submit a claim under Article 76. Over the years, however, it has submitted observations on submissions made by other states, requesting that those observations be made available online and to the commission. In addition, since 2001 the United States has gathered and analyzed data to determine the outer limits of its extended continental shelf.

Arctic border countries have begun the complex investigations needed to support claims for an extended continental shelf in the Arctic. Claims have already been submitted by several countries, including the Russian Federation, which submitted its UNCLOS claim to a portion of the Arctic continental shelf in 2001.[100] Russia made claims to the Lomonosov Ridge, an undersea feature spanning the Arctic from Russia to Canada, as an extension of its continental margin. The submission demonstrated Russia's bid to extend activities in Arctic regions. The United States has started to gather and analyze data through an initiative called the Extended Continental Shelf Project.[101] In this effort, the United States is working closely with Canada to prepare and present Canada's submission to the Commission on the Limits of the Continental Shelf.

Canada and the United States share overlapping regions of the seabed as part of the extended continental margin of both nations. Both countries have conducted research singly and jointly to map the extended continental shelf.[102] Consistent with past U.S. directives addressing the extended continental shelf and boundary issues in the Arctic,[103] the Obama Administration has undertaken an initiative to improve the delivery of relevant scientific information to officials responsible for energy development decisions in the Arctic.[104] This initiative is part of a broader response to USGS recommendations that more dialogue and collaborative science planning occur between and among the scientific community and federal agencies involved in Arctic oil and gas development decisions.[105] These USGS recommendations stem from earlier USGS findings that most of the potential oil and gas resources estimated for the Arctic are likely to exist within already agreed-upon territorial boundaries.[106] (For further discussion, see "Territorial Claims and Sovereignty Issues.")

Minerals

A warming Arctic means new opportunities and challenges for mineral exploration and development. Receding glaciers expose previously ice-covered land that could host economic mineral deposits that were previously undetectable and un-mineable below the ice. Longer summers would also extend exploration seasons even for areas not currently ice-covered but which are only accessible for ground surveys during the warmer months. In some parts of the Arctic, such as Baffin Island, Canada, less sea ice allows ships to transport heavy equipment to remote locations, and to convey ore from mines to the market further south. Some railway and mining operators are considering developing railroads and other infrastructure to transport ore year-round.[107] As with onshore oil and gas development, however, mining infrastructure that depends on footings sunk into permafrost could become unstable if the permafrost melts in response to warmer temperatures. Also, as with oil and gas development, mineral deposits that may be technically recoverable with current technology may not be economically profitable.

Some industry commentators suggest that mining might offer better long-term economic development opportunities compared to oil and gas development because of a larger permanent workforce and project lifelines of several decades.[108] Similar to oil and gas, however, industry observers note that uncertainties and knowledge gaps exist in the understanding of environmental change in the Arctic, and how to deal with the risks of significant Arctic industrial activity.[109]

One important part of the current infrastructure in the Arctic that supports oil, gas, and mineral development is the construction and use of ice roads—

built and used during the winter—but which are not passable during the warmer months. Warmer temperatures are shortening the ice road transport seasons and creating transportation challenges. For example, the opening date for tundra roads in northern Alaska usually occurred in early November prior to 1991 and has shifted to January in recent years.[110]

Oil Pollution and Pollution Response[111]

Oil Pollution Implications of Arctic Change

Climate change impacts in the Arctic, particularly the decline of sea ice and retreating glaciers, have stimulated human activities in the region, many of which have the potential to create oil pollution. A primary concern is the threat of a large oil spill in the area. Although a major oil spill has not occurred in the Arctic region,[112] recent economic activity, such as oil and gas exploration and tourism (cruise ships), increases the risk of oil pollution (and other kinds of pollution) in the Arctic. Significant spills in high northern latitudes (e.g., the 1989 *Exxon Valdez* spill in Alaska and spills in the North Sea) suggest that the "potential impacts of an Arctic spill are likely to be severe for Arctic species and ecosystems."[113]

Risk of Oil Pollution in the Arctic

A primary factor determining the risk of oil pollution in the Arctic is the level and type of human activity being conducted in the region. Although climate changes in the Arctic are expected to increase access to natural resources and shipping lanes, the region will continue to present logistical challenges that may hinder human activity in the region. For example (as discussed in another section of this report),[114] the unpredictable ice conditions may discourage trans-Arctic shipping. If trans-Arctic shipping were to occur on a frequent basis, it would represent a considerable portion of the overall risk of oil pollution in the region. In recent decades, many of the world's largest oil spills have been from oil tankers, which can carry millions of gallons of oil.[115]

Although the level of trans-Arctic shipping is uncertain, many expect oil exploration and extraction activities to intensify in the region.[116] Oil well blowouts from offshore oil extraction operations have been a source of major oil spills, eclipsing the largest tanker spills. The largest unintentional oil spill in recent history was from the 2010 *Deepwater Horizon* incident in the Gulf of Mexico.[117] During that incident, the uncontrolled well released (over an 84-day period) approximately 200 million gallons of crude oil into the Gulf.[118] The second-largest unintentional oil spill in recent history—the *IXTOC I*, estimated at 140 million gallons—was due to an oil well blowout in Mexican Gulf Coast waters in 1979.[119]

Until the 2010 *Deepwater Horizon* incident, the spill record for offshore platforms in U.S. federal waters had shown improvement from prior years.[120] A 2003 National Research Council (NRC) study of oil and gas activities on Alaska's North Slope stated "the conclusion of these analyses is that blowouts that result in large spills are unlikely."[121] Similar conclusions were made in federal agency documents regarding deepwater drilling in the Gulf of Mexico before the 2010 *Deepwater Horizon* event.[122] Some would likely contend that the underlying analyses behind these conclusions should be adjusted to account for the 2010 Gulf oil spill. However, others may argue that the proposed activities in U.S. Arctic waters present less risk of an oil well blowout than was encountered by the *Deepwater Horizon* drill rig, because the proposed U.S. Arctic operations would be in shallower waters (150 feet) than the deepwater well (approximately 5,000 feet) that was involved in the 2010 Gulf oil spill. In addition, Shell Oil has stated that the pressures in the Chukchi Sea (the location of Shell's recent interest) would be two to three times less than they were in well involved in the 2010 Gulf oil spill.[123] Regardless of these differences, even under the most stringent control systems, some oil spills and other accidents are likely to occur from equipment failure or human error.

Potential Impacts

No oil spill is entirely benign. Even a relatively minor spill, depending on the timing and location, can cause significant harm to individual organisms and entire populations. Regarding aquatic spills, marine mammals, birds, bottom-dwelling and intertidal species, and organisms in early developmental stages—eggs or larvae—are especially vulnerable. However, the effects of oil spills can vary greatly. Oil spills can cause impacts over a range of time scales, from only a few days to several years, or even decades in some cases.

Conditions in the Arctic may have implications for toxicological effects that are not yet understood. For example, oil spills on permafrost may persist in an ecosystem for relatively long periods of time, potentially harming plant life through their root systems. Moreover, little is known about the effects of oil spills on species that are unique to the Arctic, particularly, species' abilities to thrive in a cold environment and the effect temperature has on toxicity.[124]

The effects of oil spills in high latitude, cold ocean environments may last longer and cause greater damage than expected. Some recent studies have found that oil spills in lower latitudes have persisted for longer than initially expected, thus raising the concern that the persistence of oil in the Arctic may be understated. In terms of wildlife, population recovery may take longer in the Arctic because many of the species have longer life spans and reproduce at a slower rate.[125]

Response and Cleanup Challenges in the Arctic Region

Climate changes in the Arctic are expected to increase human activities in the region, many of which impose a risk of oil pollution, particularly from oil spills. Conditions in the Arctic region impose unique challenges for personnel charged with (1) oil spill response, the process of getting people and equipment to the incident, and (2) cleanup duties, either recovering the spilled oil or mitigating the contamination so that it poses less harm to the ecosystem. These challenges may play a role in the policy development for economic activities in the Arctic.

Spill Response Challenges

Response time is a critical factor for oil spill recovery. With each hour, spilled oil becomes more difficult to track, contain, and recover, particularly in icy conditions, where oil can migrate under or mix with surrounding ice.[126] Most response techniques call for quick action, which may pose logistical challenges in areas without prior staging equipment or trained response professionals. Many stakeholders are concerned about a "response gap" for oil spills in the Arctic region.[127] A response gap is a period of time in which oil spill response activities would be unsafe or infeasible. The response gap for the northern Arctic latitudes is likely to be extremely high compared to other regions.[128]

According to the former Commander of the 17th Coast Guard District (Alaska), "we are not prepared for a major oil spill [over 100,000 gallons] in the Arctic environment. The Coast Guard has no offshore response capability in Northern or Western Alaska."[129] The transportation infrastructure along Alaska's northern coast poses challenges for oil spill responders. The Coast Guard has no designated air stations north of Kodiak, AK, which is almost 1,000 miles from the northernmost point of land along the Alaskan coast in Point Barrow, AK.[130] Although some of the communities have airstrips capable of landing cargo planes, no roads connect these communities.[131] Vessel infrastructure is also limited. The nearest major port is in the Aleutian

Islands, approximately 1,300 miles from Point Barrow. A 2010 Government Accountability Office (GAO) report identified further logistical obstacles that would hinder an oil spill response in the region, including "inadequate" ocean and weather information for the Arctic and technological problems with communications.[132]

The history of oil spills and response in the Aleutian Islands raises concerns for potential spills in the Arctic region:

> The past 20 years of data on response to spills in the Aleutians has also shown that almost no oil has been recovered during events where attempts have been made by the responsible parties or government agencies, and that in many cases, weather and other conditions have prevented any response at all.[133]

Oil Spill Cleanup Challenges

The behavior of oil spills in cold and icy waters is not well understood.[134] Cleaning up oil spills in ice-covered waters will be more difficult than in other areas, primarily because effective strategies have yet to be developed. Natural oil seeps, which are a major source of oil in the arctic environment, may offer opportunities for studying the behavior of oil.[135]

The Arctic conditions present several hurdles to oil cleanup efforts. In colder water temperatures, there are fewer organisms to break down the oil through microbial degradation. Oil evaporates at a slower rate in colder water temperatures. Although slower evaporation may allow for more oil to be recovered, evaporation removes the lighter, more toxic hydrocarbons that are present in crude oil.[136] The longer the oil remains in an ecosystem, the more opportunity there is for exposure. Oil spills may get trapped in ice, evaporating only when the ice thaws. In some cases, oil could remain in the ice for years.

Icy conditions enhance emulsification—the process of forming different states of water in oil, often described as "mousse." Emulsification creates oil cleanup challenges by increasing (1) the volume of the oil/water mixture and (2) the mixture's viscosity (resistance to flow). The latter change creates particular problems for conventional removal and pumping cleanup methods.[137] Moreover, two of the major non-mechanical recovery methods—in-situ burning and dispersant application—may be limited (or "precluded") by the Arctic conditions and lack of logistical support: aircraft, vessels, and other infrastructure.[138]

Existing Policy Framework

Considering both the recent increase in human activity in the region (and expectation of further interest) and the response and recovery challenges that an oil spill would impose in Arctic waters, many would assert that the region warrants particular attention in terms of governance. However, the existing framework for international governance of maritime operations in the Arctic region lacks legally binding requirements. While the Safety of Life at Sea Convention (SOLAS) and other International Maritime Organization (IMO) conventions include provisions regarding ships in icy waters, the provisions are not specific to the polar regions. Although the IMO has "Guidelines for Ships Operating in Arctic," a 2009 NOAA report described the non-binding IMO provisions as "inconsistent with the hazards of Arctic navigation and the potential for environmental damage from such an incident."[139]

Fisheries[140]

Large commercial fisheries exist in the Arctic, including in the Barents and Norwegian Seas north of Europe, the Central North Atlantic off of Greenland and Iceland, and the Newfoundland and Labrador Seas off of northeastern Canada.[141]

In the 110th Congress, P.L. 110-243 stated that "the United States should initiate international discussions and take necessary steps with other Arctic nations to negotiate an agreement or agreements for managing migratory, transboundary, and straddling fish stocks in the Arctic Ocean and establishing a new international fisheries management organization or organizations for the region." In response to this, the United States participated in meetings with other Arctic nations at the March 2009 U.N. Food and Agriculture Organization Committee on Fisheries meetings in Rome. These international discussions are continuing. No additional legislation has been enacted related to U.S. fisheries in the Arctic.

On February 5, 2009, the North Pacific Fishery Management Council, acting under the authority of the Magnuson-Stevens Fishery Conservation and Management Act, unanimously approved the Council's draft Arctic Fisheries Management Plan and approved the preferred alternative, which would (1) close the Arctic to commercial fishing until information improves so that fishing can be conducted sustainably and with due concern to other ecosystem components; (2) determine the fishery management authorities in the Arctic and provide the Council with a vehicle for addressing future management issues; and (3) implement an ecosystem-based management policy that recognizes the resources of the U.S. Arctic and the potential for fishery development that might affect those resources, particularly in the face of a changing climate.

On November 3, 2009, the National Marine Fisheries Service (Department of Commerce) issued a final rule that implemented the North Pacific Council's Fishery Management Plan for Fish Resources of the Arctic Management Area, as well as Amendment 29 to the Fishery Management Plan for Bering Sea/Aleutian Islands King and Tanner Crabs. This action followed the North Pacific Council's recommendations establishing a basis for sustainable management of commercial fishing in the Arctic Management Area and moving the northern boundary of the crab fishery south of the Arctic Management Area to Bering Strait.[142] In October 2009, an international symposium focused on conservation of fish stocks and the potential for future fisheries management in the Arctic Ocean.[143] The National Marine Fisheries Service maintains an "Arctic Fisheries" web page summarizing activates related to fisheries.[144]

Protected Species[145]

Concern over development of the Arctic relates to how such development might affect threatened and endangered species. Under the Endangered Species Act (ESA, 16 U.S.C. §§1531-1543), the polar bear was listed as threatened on May 15, 2008. The failure by the Fish and Wildlife Service (FWS) to make a 90-day finding on a 2008 petition to list Pacific walrus led to submission of 60- days' notice of a future citizen suit. However, eventually walruses were listed as candidate species under ESA;[146] this status means that federal agencies carrying out actions that may affect the species must confer with FWS though they are not necessarily obliged to modify their actions. Both polar bears and walruses are heavily dependent during their life cycles on thick sea ice, making them especially susceptible to the shrinking Arctic ice cap.

On December 30, 2008, the National Marine Fisheries Service (NMFS) determined that a listing of ribbon seal as threatened or endangered was not warranted.[147] On October 22, 2010, NMFS listed the southern distinct population segment (DPS) of spotted seals as threatened.[148] Listing of two other DPS (Okhotsk and Bering Sea) had earlier been determined to not be warranted.[149] On December 10, 2010, NMFS proposed that (1) four subspecies of ringed seal be listed as threatened,[150] and (2) that two DPS of one subspecies of bearded seal be listed as threatened.[151]

In either terrestrial or marine environments, the extreme pace of change makes a biological response many times more difficult. For species with adaptations for a specific optimum temperature for egg development, or production of young timed to match the availability of a favored prey species, or seed dispersal in predictable fire regimes, etc., evolutionary responses may well not keep pace with the rate of change.[152] While species of plants and animals farther south might migrate, drift, or be transplanted from warming habitats to more northerly sites that may continue to be suitable,[153] once a terrestrial species reaches the Arctic Ocean, it is very literally at the end of the line. No more northern or colder habitat is available.

The Marine Mammal Protection Act (MMPA; 16 U.S.C. §§1361 et seq.) protects whales, seals, walruses, and polar bears. The MMPA established a moratorium on the "taking" of marine mammals in U.S. waters and by U.S. nationals on the high seas, including the Arctic. The MMPA protects marine mammals from "clubbing, mutilation, poisoning, capture in nets, and other human actions that lead to extinction." Under the MMPA, the Secretary of Commerce, acting through National Marine Fisheries Service (NMFS), is responsible for the conservation and management of whales and seals. The Secretary of the Interior, acting through the Fish and Wildlife Service (FWS), is responsible for walruses and polar bears.[154] Despite the MMPA's general moratorium on taking, the MMPA allows U.S. citizens to apply for and obtain authorization for taking small numbers of mammals incidental to activities other than commercial fishing (e.g., offshore oil and gas exploration and development) if the taking would have only a negligible impact on any marine mammal species or stock, provided that monitoring requirements and other conditions are met.

Indigenous People Living in the Arctic[155]

People have been living in the Arctic for thousands of years, and indigenous peoples developed highly specialized cultures and economies based on the physical and biological conditions of the long-isolated region. However, with trade, the influx of additional populations especially since the 19th century, and ongoing physical changes in the Arctic, indigenous populations have already experienced substantial change in their lifestyles and economies. Over the past two decades, greater political organization across indigenous populations has increased their demands for international recognition and broader rights, as well as attention to the economic, health, and safety implications of climate change in the North.

Background

Seven of the eight Arctic nations have indigenous peoples,[156] whose predecessors were present in parts of the Arctic over 10,000 years ago, well before the arrival of peoples with European backgrounds.[157] Current Arctic indigenous peoples comprise dozens of diverse cultures and speak dozens of languages from eight or more non-Indo-European language families.[158]

Before the arrival of Europeans, Arctic indigenous peoples lived in economies that were chiefly dependent, in varying proportions, on hunting land and marine mammals, catching salt- and fresh-water fish, herding reindeer (in Eurasia), and gathering, for their food, clothing, and other products.[159] Indigenous peoples' interaction with and knowledge of Arctic wildlife and environments has developed over millennia and is the foundation of their cultures.[160]

The length of time that Arctic indigenous peoples were in contact with Europeans varied across the Arctic. As recorded by Europeans, contact began as early as the 9th century CE, if not before, in Fennoscandia[161] and northwestern Russia, chiefly for reasons of commerce (especially furs); it progressed mostly west-to-east across northern Asia, reaching northeastern Arctic Asia by the 17th century.[162] North American Arctic indigenous peoples' contact with Europeans started in Labrador in the 16th century and in Alaska in the 18th century, and was not completed until the early 20th century.[163] Greenland's indigenous peoples first saw European-origin peoples in the late 10th century, but those Europeans died out during the 15th or 16th century and Europeans did not return permanently until the 18th century.[164]

Contact led to significant changes in Arctic indigenous economies, political structures, foods, cultures, and populations, starting especially in the 20th century. For example, life expectancy among Alaska Natives has increased from 47 years in 1950 to over 69 years in 2000 (though it still lags behind that of U.S. residents overall, at 77 years).[165]

Also, at present, most Arctic indigenous peoples have become minorities in their countries' Arctic areas, except in Greenland and Canada. (One source estimates that, around 2003, about 10% of an estimated 3.7 million people in the Arctic were indigenous.)[166] While many Arctic indigenous communities remain heavily dependent on hunting, fishing, and herding and are more likely to depend on traditional foods than non-indigenous Arctic inhabitants,[167] there is much variation. Most Arctic indigenous people may no longer consume traditional foods as their chief sources of energy and nutrition.[168] Major economic change is also relatively recent but ongoing.[169] Many Arctic indigenous communities have developed a mixture of traditional economic activities and wage employment.[170] The economics of subsistence and globalization will be key factors in the effects of climate change on Arctic indigenous peoples, and on their reactions to Arctic climate change.

Arctic indigenous peoples' current political structures vary, as do their relationships with their national governments. Some indigenous groups govern their own unique land areas within the national structure, as in the United States and Canada; others have special representative bodies, such as the Saami parliaments in Norway, Finland, and Sweden;[171] a few areas have general governments with indigenous majorities, such as Greenland (a member country of Denmark), Nunavut territory in Canada, and the North Slope and Northwest Arctic boroughs in Alaska.[172] Control of land, through claims and ownership, also varies among Arctic indigenous peoples, as do rights to fishing, hunting, and resources.[173] Arctic indigenous peoples' political relationships to their national and local governments, and their ownership or claims regarding land, are also significant factors in the responses to Arctic climate change by the indigenous peoples and by Arctic nations' governments.

Effects of Climate Change

Arctic climate change is expected to affect the economies, population, subsistence, health, infrastructure, societies, and cultures of Arctic indigenous peoples. Changes in sea ice and sea level, permafrost, tundra, weather, and vegetation distributions, as well as increased commercial shipping, mineral extraction, and tourism, will affect the distribution of land and sea mammals, of freshwater and marine fish, and of forage for reindeer. These will in turn affect traditional subsistence activities and related indigenous lifestyles.[174] Arctic indigenous peoples' harvesting of animals is likely to become riskier and less predictable, which may increase food insecurity, change diets, and increase dependency on outside, non-traditional foods.[175] Food cellars in many locations have thawed during summers, threatening food safety. Related health risks of diabetes, obesity, and mental illness have been associated with these changes.[176]

Sea, shoreline ice, and permafrost changes have damaged infrastructure and increased coastal and inland erosion, especially in Alaska, where GAO found in 2003 that "coastal villages are becoming more susceptible to flooding and erosion caused in part by rising temperatures."[177] In response, Congress funded the U.S. Army Corps of Engineers to conduct a Baseline Erosion Assessment that identified and prioritized among the 178 communities identified at risk from erosion.[178] (Risks from flooding were not examined.) GAO concluded in 2009 that many Native villages must relocate, but even those facing imminent threats have been impeded by various barriers, including difficulties identifying appropriate new sites, piecemeal programs for state and federal assistance, and obstacles to eligibility for certain federal programs.[179] The Alaska Federation of Natives placed among its 2010 federal priorities a request to Congress to mitigate flooding and erosion in Alaska Native villages and to fund relocation of villages where necessary.[180] However, "the cost is extraordinary," acknowledges Senator Lisa Murkowski.[181]

Oil, gas, and mineral exploration and development are expected to increase, as are other economic activities, such as forestry and tourism, and these are expected to increase economic opportunities for all Arctic residents, including indigenous peoples.[182] Pressures to increase participation in the wage economy, however, may speed up changes in indigenous cultures. Increased economic opportunities may also lead to a rise in the non-indigenous population, which may further change the circumstances of indigenous cultures. Some representatives of Arctic indigenous people have related a "conflicting desire between combating climate change and embracing the potential for economic growth through foreign investment."[183]

Although important advances in public health have occurred in indigenous communities over past decades, some health problems may increase with continued Arctic climate change. Economic development may exacerbate Arctic pollution problems, including higher exposure to mercury, air pollution, and food contamination. The influx and redistribution of contaminants in the air, oceans, and land may change in ways that are now poorly understood.[184] Warmer temperatures and longer warm seasons may increase insect- and wildlife-borne diseases.[185] Climate change may lead to damage to water and sanitation systems, reducing protection against waterborne diseases.[186] Changes in Arctic indigenous cultures may increase mental stress and behavioral problems.[187]

The response to climate change by Arctic indigenous peoples has included international activities by Arctic indigenous organizations and advocacy before their national governments. As one report noted, "the rise of solidarity among indigenous peoples organizations in the region is surely a development to be reckoned with by all those interested in policy issues in the Arctic."[188] Six national or international indigenous organizations are permanent participants of the Arctic Council, the regional intergovernmental forum.[189] Due in part to advocacy by Arctic indigenous people, the United Nations General Assembly adopted in 2007 the Declaration on the Rights of Indigenous Peoples.[190] In April 2009, the Inuit Circumpolar Council (an organization of Inuit in the Arctic regions of Alaska, Canada, Greenland, and Russia) hosted in Alaska the worldwide "Indigenous Peoples Global Summit on Climate Change."[191] The conference report, forwarded to the Copenhagen Conference of the Parties of the U.N. Framework Convention on Climate Change (December 2009), noted "accelerating" climate change caused by "unsustainable development" and, among several recommendations, called for a greater indigenous role in national and international decisions on climate change, including a greater role for indigenous knowledge in climate change research, monitoring, and mitigation.[192]

Polar Icebreaking[193]

The Coast Guard's polar icebreakers perform a variety of missions supporting U.S. interests in the Arctic (and Antarctic), including the following:

- conducting and supporting U.S. scientific research in the Arctic;[194]
- defending U.S. sovereignty in the Arctic by helping to maintain a presence in the region;
- defending other U.S. interests in the Arctic, including economic interests relating to the U.S. exclusive economic zone (EEZ) north of Alaska;
- monitoring sea traffic in the Arctic, including ships bound for the United States; and
- conducting other typical Coast Guard missions (such as search and rescue, law enforcement, and protection of marine resources) in Arctic waters, including U.S. territorial waters north of Alaska.

The Coast Guard's strategy document for the Arctic region, released on May 21, 2013, states that "The United States must have adequate icebreaking capability to support research that advances fundamental understanding of the region and its evolution," and that "The Nation must also make a strategic investment in icebreaking capability to enable access to the high latitudes over the long-term."[195]

The Coast Guard's two existing heavy polar icebreakers—*Polar Star* and *Polar Sea*—have exceeded their originally intended 30-year service lives. *Polar Star* was placed in caretaker status on July 1, 2006. Congress in FY2009 and FY2010 provided funding to repair it and return it to service for an additional 7 to 10 years of service; the repair work was completed and the ship was reactivated on December 14, 2012.

On June 25, 2010, the Coast Guard announced that *Polar Sea* had suffered an unexpected engine casualty; the ship was unavailable for operation after that. The Coast Guard placed *Polar Sea* in commissioned, inactive status on October 14, 2011. Section 222 of the Coast Guard and Maritime Transportation Act of 2012 (H.R. 2838/P.L. 112-213 of December 20, 2012) prohibits the Coast Guard from removing any part of *Polar Sea* and from transferring, relinquishing ownership of, dismantling, or recycling the ship until it submits a business case analysis of the options for and costs of reactivating the ship and extending its service life to at least September 30, 2022, so as to maintain U.S. polar icebreaking capabilities and fulfill the Coast Guard's high latitude mission needs, as identified in the Coast Guard's July 2010 High Latitude Study.

The Coast Guard's third polar icebreaker—*Healy*—entered service in 2000. Compared to *Polar Star* and *Polar Sea*, *Healy* has less icebreaking capability (it is considered a medium polar icebreaker), but more capability for supporting scientific research. The ship is used primarily for supporting scientific research in the Arctic.

The Coast Guard's FY2013 budget initiated a new project for the design and construction of a new polar icebreaker. The Coast Guard's proposed FY2013 budget requested $8 million in FY2013 acquisition funding to initiate survey and design activities for the ship, and projected an additional $852 million in FY2013-FY2017 for acquiring the ship. The Coast Guard's FY2013 budget anticipated awarding a construction contract for the ship "within the next five years" and taking delivery on the ship "within a decade."

The Coast Guard's proposed FY2014 budget requests $2 million in acquisition funding to continue survey and design activities for the ship, or $118 million less than the $120 million that was projected for FY2014 under the FY2013 budget. The Coast Guard's FY2014 budget submission projects an additional $228 million in FY2015-FY2018 for acquiring the ship, including $128 million in FY2015-FY2017, or $604 million less than the $732 million that was projected for FY2015-FY2017 under the Coast Guard's FY2013 budget submission. The Coast Guard's proposed FY2014 budget anticipates awarding a construction contract for the ship "within the next four years."

The Coast Guard states that the reduction in funding for the program in the FY2014 budget compared to the FY2013 budget does not materially change the schedule for acquiring the ship, for two reasons. First, the Coast Guard states that the prohibition on new program starts that was in place under the continuing resolution (CR) that funded government operations from October 1, 2012, until March 26, 2013 (H.J.Res. 117/P.L. 112-175 of September 28, 2012) will prevent the Coast Guard from fully using the $8 million in FY2013 funding that was provided for the program, permitting the unused portion to be carried over into FY2014. This carried-over funding, combined with the $2 million requested for FY2014, will be enough, the Coast Guard states, to fund FY2014 activities for the program. Second, and more generally, the Coast Guard states that, compared to last year, the Coast Guard now has a more fully developed understanding of the schedule for the polar icebreaker project, and that much of the funding for the ship that was included in the FY2013 CIP can now be viewed in retrospect as having been put into the plan ahead of need.[196]

With the reactivation of *Polar Star*, the operational U.S. polar icebreaking fleet consists of one heavy polar icebreaker (*Polar Star*) and one medium polar icebreaker (*Healy*). The new polar icebreaker for which initial acquisition funding is requested in the FY2013 budget would replace *Polar Star* at about the time Polar Star's 7- to 10-year reactivation period ends.

In July 2011, the Coast Guard provided to Congress a study on the Coast Guard's missions and capabilities for operations in high-latitude (i.e., polar) areas. The study, commonly known as the High Latitude Study and dated July 2010 on its cover, concluded the following:

- **The Coast Guard requires three heavy and three medium icebreakers to fulfill its statutory missions.** These icebreakers are necessary to (1) satisfy Arctic winter and transition season demands and (2) provide sufficient capacity to also execute summer missions. Single-crewed icebreakers have sufficient capacity for all current and expected statutory missions. Multiple crewing provides no advantage because the number of icebreakers required is driven by winter and shoulder season requirements. Future use of multiple or augmented crews could provide additional capacity needed to absorb mission growth.
- **The Coast Guard requires six heavy and four medium icebreakers to fulfill its statutory missions and maintain the continuous presence requirements of the Naval Operations Concept.** Consistent with current practice, these icebreakers are single-crewed and homeported in Seattle Washington.

- **Applying crewing and home porting alternatives reduces the overall requirement to four heavy and two medium icebreakers.** This assessment of non-material solutions shows that the reduced number of icebreakers can be achieved by having all vessels operate with multiple crews and two of the heavy icebreakers homeporting in the Southern Hemisphere.

> Leasing was also considered as a nonmaterial solution. While there is no dispute that the Coast Guard's polar icebreaker fleet is in need of recapitalization, the decision to acquire this capability through purchase of new vessels, reconstruction of existing ships, or commercial lease of suitable vessels must be resolved to provide the best value to the taxpayer. The multi-mission nature of the Coast Guard may provide opportunities to conduct some subset of its missions with non government-owned vessels. However, serious consideration must be given to the fact that the inherently governmental missions of the Coast Guard must be performed using government-owned and operated vessels. An interpretation of the national policy is needed to determine the resource level that best supports the nation's interests....
>
> The existing icebreaker capacity, two inoperative heavy icebreakers and an operational medium icebreaker, does not represent a viable capability to the federal government. The time needed to augment this capability is on the order of 10 years. At that point, around 2020, the heavy icebreaking capability bridging strategy expires.[197]

At a July 27, 2011, hearing on U.S. economic interests in the Arctic before the Oceans, Atmosphere, Fisheries, and Coast Guard subcommittee of the Senate Commerce, Science, and Transportation Committee, the following exchange occurred:

> SENATOR OLYMPIA J. SNOWE: On the high latitude study, do you agree with—and those—I would like to also hear from you, Admiral Titley, as well, on these requirements in terms of Coast Guard vessels as I understand it, they want to have—I guess, it was a three medium ice breakers. Am in correct in saying that? Three medium ice breakers.
>
> ADMIRAL ROBERT PAPP, COMMANDANT OF THE COAST GUARD: I agree with the mission analysis and as you look at the requirements for the things that we might do up there, if it is in the nation's interest, it identifies a minimum requirement for three heavy ice breakers and three medium ice breakers and then if you want a persistent presence up there, it would require—and also doing things such as breaking out (inaudible) and other responsibilities, then it would take up to a maximum six heavy and four medium.
>
> SNOWE: Right. Do you agree with that?

PAPP: If we were to be charged with carrying out those full responsibilities, yes, ma'am. Those are the numbers that you would need to do it.

SNOWE: Admiral Titley, how would you respond to the high latitude study and has the Navy conducted its own assessment of its capability?

REAR ADMIRAL DAVID TITLEY, OCEANORGRAPHER AND NAVIGATOR OF THE NAVY: Ma'am, we are in the process right now of conducting what we call a capabilities based assessment that will be out in the summer of this year.

We are getting ready to finish that—the Coast Guard has been a key component of the Navy's task force on climate change, literally since day one when the Chief of Naval Operations set this up, that morning, we had the Coast Guard invited as a member of our executive steering committee.

So we have been working very closely with the Coast Guard, with the Department of Homeland Security, and I think Admiral Papp—said it best as far as the specific comments on the high latitude study but we have been working very closely with the Coast Guard.[198]

Potential issues for Congress regarding Coast Guard polar icebreaker modernization include the following:

- the numbers and capabilities of polar icebreakers the Coast Guard will need in the future;
- the disposition of *Polar Sea*;
- whether the new polar icebreaker initiated in the FY2013 budget should be funded with incremental funding (as proposed in the Coast Guard's Five Year Capital Investment Plan) or full funding in a single year, as required under the executive branch's full funding policy;
- whether new polar icebreakers should be funded entirely in the Coast Guard budget, or partly or entirely in some other part of the federal budget, such as the Department of Defense (DOD) budget, the National Science Foundation (NSF) budget, or both;
- whether to provide future icebreaking capability through construction of new ships or service life extensions of existing polar icebreakers; and whether future polar icebreakers should be acquired through a traditional acquisition or a leasing arrangement.

Search and Rescue[199]

General

The possibility of increased sea and air traffic through Arctic waters has increased concerns regarding Arctic-area search and rescue capabilities. Given the location of current U.S. Coast Guard operating bases, it could take Coast Guard aircraft several hours, and Coast Guard cutters days or even weeks, to reach a ship or a downed aircraft in distress in Arctic waters. In addition, the harsh climate complicates search and rescue operations in the region. Particular concern has been expressed about cruise ships that may experience problems and need assistance; there have already been incidents of this kind in recent years in waters off Antarctica. A May 7, 2011, press report stated:

> U.S. and Canadian military commanders say they are examining their rescue capabilities in the Arctic as a shrinking ice cap brought about by climate change opens up rich oil and gas reserves and draws more commercial traffic to the top of the globe....
>
> [U.S. Admiral James Winnefeld, commander of the U.S. Northern Command and the North American Aerospace Defense Command, or NORAD] and NORAD'S deputy commander, Canadian Forces Lt. Gen. Marcel Duval, said more ships in the Arctic Ocean - and more jetliners crossing Arctic skies - could mean more emergencies.
>
> "More ships, more chances of accidents," Duval said in a separate interview in his office at NORAD headquarters at Peterson.
>
> Last summer, the Canadian Coast Guard rescued 197 people from the cruise ship Clipper Adventurer, which ran aground inside the Arctic Circle....
>
> Sea traffic is still light, with only about 25 ships a year currently crossing the maritime Arctic boundary between Alaska and the Yukon. But that number is increasing by 10 to 15 percent a year, according to NORAD statistics.
>
> Civilian air traffic over the Arctic is booming. U.S. and Canadian aviation agencies report more than 9,600 civilian flights across the North Pole in 2010, up nearly 21 percent from 2008.[200]

One option for mitigating the risks associated with cruise ship operations in the Arctic would be to require the vessels to sail in pairs, so that if one ship experiences a problem, the other one could quickly come to its aid.

Increasing U.S. Coast Guard search and rescue capabilities for the Arctic could require one or more of the following: enhancing or creating new Coast Guard operating bases in the region; procuring additional Arctic-capable aircraft, cutters, and rescue boats for the Coast Guard; and adding systems to improve Arctic maritime communications, navigation, and domain awareness.[201] It may also entail enhanced forms of cooperation with navies and coast guards of other Arctic countries.

A May 13, 2013, press report states:

> Arctic nations must urgently improve rescue services in the resource-rich region that is opening up fast to shipping, energy and mining companies, Canadian experts said on Monday [May 13]....
>
> "It is crucial that northern nations strengthen response capabilities to shipping-related accidents foreseen in newly-opened northern waters," scientists from the Canadian Munk- Gordon Arctic Security Program wrote in a report.
>
> Faster responses to emergencies for people living in the Arctic, including floods or forest fires, were also urgently need[ed], the report said.[202]

Coast Guard Statements

A January 2011 press report summarized remarks made by the Commandant of the Coast Guard, Admiral Robert Papp, concerning Coast Guard search and rescue capabilities in the Arctic. The article quoted Papp as saying that there is a need for a range of Coast Guard operational capability for the Arctic, and that "In the meantime, he said the service will lean on partnerships with other Arctic nations. However, he warned that the Coast Guard will likely not be able to respond to any crises in the Arctic circle in a timely fashion. He recalled that the Canadian Coast Guard came under fire when it took six days to rescue a cruise ship and oil tanker that both ran aground in its northern waters." The article stated:

> "We wouldn't be able to make it in six days," he said. "It'd probably take us six weeks to get adequate resources up for a similar thing in our waters, so we have to start focusing on this."[203]

A March 2011 press report summarizing remarks made by Admiral Papp during and after a hearing before the Coast Guard and Maritime Transportation subcommittee of the House Transportation and Infrastructure Committee stated:

> Because retreating ice continues to make the passageways in the Arctic Ocean more navigable, the Coast Guard needs to have air stations in the region to conduct helicopter rescue missions, Papp said. He has not

chosen specific cities for the air station sites, but he said the Alaskan cities of Barrow and Kotzebue come to mind.

When lawmakers focus on Arctic resources, the United States' shortage of icebreakers usually draws the bulk of attention, but Papp said the Coast Guard is also in dire need of small boats to conduct rescue missions.[204]

At a June 23, 2011, hearing on the Coast Guard's proposed FY2012 budget before the Oceans, Atmosphere, Fisheries, and Coast Guard subcommittee of the Senate Commerce, Science, and Transportation Committee, the following exchange occurred:

ADMIRAL ROBERT J. PAPP, JR., COMMANDANT OF THE COAST GUARD: I visited Barrow, Kotzebue, and Nome. And, actually, it was a revisit because I had served up there as an ensign (ph) 35 years ago. And so it was good to get back up there and see the changes.

But what has not changed is the infrastructure up there. And I think that we have to have a robust discussion on the infrastructure needed to support what is no doubt going to be an increase in human activity up there off the – the North Coast of Alaska.

Icebreakers I think are important, but they cloud the discussion of the other needs that we have up there. And I think we've focused on icebreakers over the last few years, even though they are important...

SENATOR MARK BEGICH: Right.

PAPP: ... and needed. But, right now, if we were to have to mount a response like we did in the Gulf of Mexico—I sent 3,000 people down for Deepwater Horizon. You know how many hotel rooms are available in Barrow.

BEGICH: That's right.

PAPP: And we—we have no place to put people up there. We have no hangars for aircraft. We have no piers, no Coast Guard boats. So my immediate, pressing concern is as human activity occurs, as you have that ship that goes through that—first of all, we'll assure safety standards because no matter where ships operate in the world, we are involved in their safety standards.

But if an accident happens, how do we respond? And, right now, we've got zero capability to respond in the Arctic right now. And we've got to do better than that.

That—when people ask me what keeps me awake at night—an oil spill, a collision, a ship sinking in the Arctic keeps me awake at night because we have nothing to respond or, if we respond, it's going to take us weeks to get there.[205]

State Department Statement

The State Department states that

> the United States and the seven other Arctic Council (AC) Member States (Canada, Denmark, Finland, Iceland, Norway, Russian Federation, Sweden) adopted a Ministerial Declaration in April 2009 in Tromso, Norway, which established a Task Force with the mandate of developing a Search and Rescue (SAR) agreement for the Arctic. The United States hosted the first meeting of the Arctic SAR Task Force during December 9-11, 2009, in Washington, D.C. The next round of negotiations is scheduled [sic] for February 25-26, 2010, in Moscow, Russian Federation. The Arctic SAR Task Force has been asked to finalize the Arctic SAR agreement such that it can be presented for adoption by the AC at its Ministerial meeting in Spring 2011.
>
> The United States is at the forefront of efforts to promote safety in the Arctic. The Department of State's Office of Ocean and Polar Affairs is coordinating federal interagency interest to negotiate an instrument for the saving of lives at sea and the rescue of survivors after aircraft accidents in the Arctic. The U.S. Coast Guard is a prominent agency participating in this effort. With so few resources available for SAR in the Arctic, developing a regional agreement to set baseline standards for greater international cooperation and coordination will be of great value.[206]

Arctic Council Agreement on Arctic Search and Rescue (May 2011)

On May 12, 2011, representatives from the member states of the Arctic Council, meeting in Nuuk, Greenland, signed an agreement on cooperation on aeronautical and maritime search and rescue in the Arctic. Secretary of State Hillary Rodham Clinton signed for the United States. Key features of the agreement include the following:

- Article 2 states: "The objective of this Agreement is to strengthen aeronautical and maritime search and rescue cooperation and coordination in the Arctic."
- Article 3 and the associated Annex to the agreement essentially divide the Arctic into search and rescue areas within which each party has primary responsibility for conducting search and rescue operations, stating that "the delimitation of search and rescue regions is not related to and shall not prejudice the delimitation of any boundary between States or their sovereignty, sovereign rights or jurisdiction," and that "each Party shall promote the establishment, operation and

maintenance of an adequate and effective search and rescue capability within its area."

- Article 4 and the associated Appendix I to the agreement identify the competent authority for each party. For the United States, the competent authority is the Coast Guard.
- Article 5 and the associated Appendix II to the agreement identify the agencies responsible for aeronautical and maritime search and rescue for each party. For the United States, those agencies are the Coast Guard and the Department of Defense.
- Article 6 and the associated Appendix III to the agreement identify the aeronautical and/or maritime rescue coordination centers (RCCs) for each party. For the United States, the RCCs are Joint Rescue Coordination Center Juneau (JRCC Juneau) and Aviation Rescue Coordination Center Elmendorf (ARCC Elmendorf).
- Article 7 states that "the Parties shall conduct aeronautical and maritime search and rescue operations pursuant to this Agreement consistent" with certain general guidelines.[207]
- Article 8 states that "a Party requesting permission to enter the territory of a Party or Parties for search and rescue purposes, including for refueling, shall send its request to a search and rescue agency and/or RCC of the relevant Party or Parties," and that "the Party receiving such a request shall immediately confirm such receipt. The receiving Party, through its RCCs, shall advise as soon as possible as to whether entry into its territory has been permitted and the conditions, if any, under which the mission may be undertaken."
- Article 9 states that "the Parties shall enhance cooperation among themselves in matters relevant to this Agreement," that "the Parties shall exchange information that may serve to improve the effectiveness of search and rescue operations," and that "the Parties shall promote mutual search and rescue cooperation by giving due consideration to collaborative efforts."
- Article 10 states that "the Parties shall meet on a regular basis in order to consider and resolve issues regarding practical cooperation."
- Article 11 states that "after a major joint search and rescue operation, the search and rescue agencies of the Parties may conduct a joint review of the operation led by the Party that coordinated the operation."
- Article 12 states that "unless otherwise agreed, each Party shall bear its own costs deriving from its implementation of this Agreement,"

and that "implementation of this Agreement shall be subject to the availability of relevant resources."

- Article 18 states that "any Party to this Agreement may, where appropriate, seek cooperation with States not party to this Agreement that may be able to contribute to the conduct of search and rescue operations, consistent with existing international agreements."
- Article 19 states that "any Party may at any time withdraw from this Agreement by sending written notification thereof to the depositary[208] through diplomatic channels at least six months in advance."[209]

Figure 4 shows an illustrative map of the national areas of search and rescue responsibility based on the geographic coordinates listed in the Annex to the agreement.

The State Department states that the agreement:

> is the first legally-binding instrument negotiated under the auspices of the Arctic Council. It coordinates life-saving international maritime and aeronautical SAR coverage and response among the Arctic States across an area of about 13 million square miles in the Arctic.
>
> As Arctic sea ice coverage decreases, ship-borne activities are increasing significantly in the Arctic. Flight traffic is also on the rise as new polar aviation routes cross the Arctic air space in several directions. As human presence and activities in the Arctic expand, the potential for accidents increases as well. Limited rescue resources, challenging weather conditions, and the remoteness of the area render SAR operations difficult in the Arctic, making coordination among the Arctic nations imperative. The SAR Agreement will improve search and rescue response in the Arctic by committing all Parties to coordinate appropriate assistance to those in distress and to cooperate with each other in undertaking SAR operations. For each Party, the Agreement defines an area of the Arctic in which it will have lead responsibility in organizing responses to SAR incidents, both large and small. Parties to the Agreement commit to provide SAR assistance regardless of the nationality or status of persons who may need it.
>
> The Arctic Council launched this initiative at its 2009 Ministerial Meeting in Tromso, Norway, establishing a Task Force, co-chaired by the United States and the Russian Federation. The Task Force proceeded in a highly collaborative spirit, meeting five times (in Washington, Moscow, Oslo, Helsinki and Reykjavik)
>
> The signature of the SAR Agreement in Nuuk is a positive step toward building partnerships in the Arctic. In particular, it reflects the commitment of the Arctic Council States to enhance their cooperation

and offer responsible assistance to those involved in accidents in one of the harshest environments on Earth.

(Based on geographic coordinates listed in the agreement)

Source: "Arctic Search and Rescue Agreement," accessed July 7, 2011, at http://www.arcticportal.org/features/features-of-2011/arctic-search-and-rescue-agreement.

Figure 4. Illustrative Map of Arctic SAR Areas in Arctic SAR Agreement.

This Agreement illustrates one of the most successful negotiations to date to address emerging issues in the Arctic. Arctic Council participants approached SAR negotiations with collaboration and dedication to a positive outcome. The United States congratulates its colleagues in this effort and looks forward to further collaboration on the vital issues facing the rich but fragile Arctic region.[210]

Geopolitical Environment[211]

Many observers have noted that the loss of Arctic ice[212] is already leading to stepped-up human activity in the high north, particularly in the form of increasing commercial traffic and economic development. This trend has brought forth a range of issues on the geopolitical front, from environmental protection to search-and-rescue capabilities to the delineation of national boundaries—which will determine access to natural resources. These concerns are being addressed cooperatively in both bilateral and multilateral fashion, especially under the aegis of the Arctic Council and the U.N. Convention on the Law of the Sea (UNCLOS). International law Professor William Moomaw of the Fletcher School of Law and Diplomacy has noted that "the lure of riches in the Arctic draws ever more companies and nations, but so far it's been relatively amicable jousting and jostling there."[213]

Nonetheless, some observers continue to raise questions concerning security in the high north, and have advanced sometimes conflicting views regarding the potential roles of military forces in the region. Of the other Arctic coastal nations, the United States enjoys strong political and economic ties with Canada, Norway, and Denmark; all four countries are members of NATO. Although the United States views Russia as an important partner in developing policies to cope with changing conditions in the Arctic, relations with Moscow have had numerous areas of tension in recent years. Several non-Arctic nations, including India and China, have also evinced interest in the high north, and recently gained permanent observer status in the Arctic Council. In addition, the European Union, which sought but was denied full observer status in 2009 and 2013, is nonetheless developing policy toward the Arctic.[214]

Multilateral Cooperation

As noted elsewhere in this report, in late May 2008, ministerial representatives of the five Arctic littoral states attended a meeting convened by Denmark in Ilulissat, Greenland (a semi- autonomous territory of Denmark). Danish Foreign Minister Stig Møller implied that the meeting was intended to develop interim measures for Arctic governance: "We must continue to fulfill our obligations in the Arctic area until the U.N. decides who will have the right to the sea and the resources in the region. We must agree on the rules and what to do if climate changes make more shipping possible."[215] Attendees discussed a variety of issues, including the environment, transportation, resources, and security. The Danish Foreign Ministry stated that the resulting Ilulissat Declaration

> sent a clear political signal to the local inhabitants and the rest of the world that we will act responsibly when addressing the development in the Arctic Ocean. We have committed ourselves politically to solve any

> disagreements through negotiation. Thus, hopefully, we have eradicated all the myths about a 'race for the North Pole.' The legal framework is in place and the five States have now declared that they will abide by it.[216]

The Arctic has increasingly become a subject of discussions in bilateral meetings among leaders of the nations in the region and elsewhere around the globe. The main international forum for cooperation in the high north, however, is the eight-nation Arctic Council, of which the United States is an active member. The chief topics addressed by the six working groups of the Council, which was formed in 1996, are sustainable development, environmental protection, and the social well-being of the indigenous communities. The United States vetoed security as an issue of consideration for the Council. The Council holds ministerial-level meetings biennially, while the working groups meet more frequently. Sweden recently held the two-year revolving Council chairmanship; Canada took over in May 2013, and will be followed by the United States in 2015.[217]

In May 2011, former Secretary of State Hillary Clinton attended the Arctic Council ministerial summit, held in Nuuk, Greenland; she was accompanied by U.S. Interior Secretary Ken Salazar. They were the first U.S. Cabinet members to attend an Arctic Council meeting, and observers stated that their visits served to raise the profile of Arctic issues. Noting the increased commercial activity in the region, Secretary Clinton declared, "We need to pursue these opportunities in a smart, sustainable way that preserves the Arctic environment and ecosystem."[218] Among other issues, attendees focused on efforts to reduce emissions that cause "black carbon" to settle on the Arctic region, accelerating ice melt. In addition, the Council discussed launching a longer-term study on methods to cope with possible future oil spills.

The major "deliverable" of the summit, however, was the signing of an Agreement on Cooperation on Aeronautical and Maritime Search and Rescue (SAR) in the Arctic. The accord was hailed as the "first legally-binding instrument negotiated under the auspices of the Arctic Council." The SAR initiative, developed mainly by the United States and Russia, had been introduced during the April 2009 summit. The Council also approved the establishment of a permanent secretariat, to be based in Tromsø, Norway.[219]

In early June 2012, Secretary Clinton traveled once more to the Arctic, visiting the newly opened secretariat in Tromsø as part of an eight-day trip to Scandinavia. While there, she emphasized that the U.S. government "want[s] the Arctic Council to remain the premier institution that deals with Arctic questions."[220]

The Council held its most recent summit on May 15, 2013 in Kiruna, Sweden. Secretary of State John Kerry's attendance, following on the heels of Secretary Clinton's two years earlier, underlined the growing importance attached by the U.S. government to polar issues. The Council revisited several of its core topics, including ocean acidification, ice loss, black carbon, biodiversity, the status of indigenous peoples, and emergency preparedness. Consideration of the latter issue resulted in the approval of a second Council accord: the Agreement on Cooperation on Marine Oil Pollution Preparedness and Response in the Arctic.

From a geopolitical perspective, however, perhaps the most important step taken by the Council was enlargement; six countries—China, India, Singapore, South Korea, Japan, and Italy—were approved for permanent observer status.[221] Prior to the meeting, the five Nordic countries reportedly had endorsed the admission of new observers, while Russia and Canada were said to be opposed—the latter for fear of diluting the interest of indigenous peoples. Washington played its cards close to the vest, remaining silent on the issue until the day of the decision. Kerry reportedly brokered the compromise. Not long before the conference, one analyst summarized the arguments for and against adding new seats to the table:

> Openness would be a shrewd move, at a stroke enhancing the council's legitimacy and the quality of its deliberations, and reducing the risk of being bypassed by countries acting unilaterally. Some of the council's recent applicants could also represent a financial boon: China, South Korea, India and Japan have the fastest-growing stable of Arctic scientists, while the EU last year proposed devoting €80 billion to Arctic research. Meanwhile, the risks of admitting new members, even heavy-hitters like China, are low. First, observers' powers are limited: they may engage with the forum's working groups, propose projects and state their views (all with permission), but they have no vote. Secondly, observer status does not alter international law: under UNCLOS, extra-regional actors have no jurisdiction in Arctic waters, and no applicant disputes this. Thirdly, the council is still fundamentally devoted to promoting research and knowledgesharing, and new observers could contribute greatly in this regard. Some Permanent Participants also fear being marginalised if new observers are admitted, but the same countries could still engage bilaterally with Arctic states from outside of the council, and there is no difference in power between permanent and ad hoc observer status. The main danger of admitting too many new observers is therefore likely to be confined to a diminished capacity for reaching swift consensus.[222]

According to some observers, the decision to add India and Singapore—which are situated thousands of miles from the Arctic Circle—would seem to indicate that the Council is transforming itself from a regional to an international forum. The revolving chair of the Council now passes from Scandinavia to North America—Canada will hold the post until 2015, when the United States will take over. Canadian Senior Arctic Official Patrick Borbey stated that his government hopes the Council will "evolve from a very solid organization doing great scientific assessment to one that actually makes policy and implements it ... and makes sure it's monitored on an ongoing basis."[223]

It has been noted that Russia "has at least half of the Arctic in terms of area, coastline, population and probably mineral wealth."[224] Moscow is keen to capitalize on natural resource development and shipping in the region. As noted elsewhere in this report, Russia and Norway in late 2010 resolved a 40-year dispute over national borders in the Barents Sea; the accord permits exploration for undersea oil, believed to be in rich supply there. In March 2013, it was announced that Russia and China had signed an agreement under which China would purchase oil from Russia in exchange for exploration licenses in the Arctic. In addition, scientists estimate that the sea route along the Siberian coast (referred to as the Northeast Passage or the Northern Sea Route, or NSR) will be ice-free and navigable well before the Northwest Passage through the Canadian archipelago. Indeed, the route is already being used: 34 vessels traversed the NSR in 2011, while 46 sailed the passage in 2012. Russia sees significant economic opportunities in offering icebreaker escorts, refueling posts, and supplies to the commercial ships that will ply the waterway.[225] The NSR will cut transport times and costs for Russian ships as well—the development of the shale gas industry in the United States means that Russia will likely now turn to Asia to market its liquefied natural gas (LNG). But Russian leaders also understand that capitalizing on new opportunities will require international cooperation and goodwill. In August 2012, Russia joined the United States and Norway in the Barents Sea for the *Northern Eagle 2012* naval maneuvers, an exercise that Norwegian Defense Minister Espen Barth Eide referred to as "a strong and interest-based neighborhood cooperation with Russia."[226]

Although its borders lie some distance from the Arctic, China has also displayed a growing interest in the region, based mainly upon the potential opportunities for shorter sea routes and the eventual development of energy-related natural resources, as well as metals and minerals.[227] China's economy is strongly dependent upon exports; some analysts have estimated that as much as one-half of China's GDP is reliant upon exports and shipping. Ocean transportation is the chief avenue for China's large petroleum imports. Beijing is keenly interested in having free access to the future northern waterways, which would drastically reduce both sailing times and transportation costs. The Chinese have announced that they may conduct their first commercial trans-Arctic voyage in the summer of 2013. Other observers have argued that China's interest is motivated in the Arctic's emerging status as "the new fishing grounds—the world's largest storehouse of biological protein."[228] Finally, some have also interpreted China's growing interest in the Arctic as further evidence of its effort to expand its influence as a global player: "They know that [the] Arctic may be one of the hot spots of the 21st century."[229]

Many Chinese believe that the Arctic should be considered as part of the "global commons." According to David Curtis Wright of the U.S. Naval War College, "The mantra that the Arctic and its natural resource wealth belong to no one country but constitute the common heritage of all humankind is virtually *de rigueur* in recent Chinese public commentary on Arctic affairs."[230] This attitude was reflected during a May 2012 workshop on Sino-Nordic cooperation hosted in Beijing, when Chinese participants referred to their country as a "near-Arctic state" and a "stakeholder." Some analysts believe that China will likely remain officially circumspect on this question, as its "foreign policy rests on a profound respect for territorial integrity."[231]

Nonetheless, some Chinese analysts reportedly are encouraging their government to challenge Canada's claim of sovereignty over the Northwest Passage.

In April 2012, former Chinese Premier Wen Jiabao visited Sweden and Iceland, and two months later former President Hu Jintao went to Denmark; the two leaders reportedly discussed large- scale investments in the region. In April 2013, China and Iceland signed a free trade agreement – China's first such pact with a European government.[232] In addition, China (like several other nations) has established a research station in the Svalbard archipelago and has beefed up the size of its embassy staff in Iceland. China has been active in conducting research on the Arctic; it acquired a large icebreaker in 1993 and has constructed a state-of-the-art polar capable research vessel, the *Snow Dragon*; in August 2012, the vessel conducted a trans-Arctic voyage, from Shanghai to Iceland; it was China's fifth Arctic research expedition.[233]

The attendance by Secretaries of State Clinton and Kerry of the last two Arctic Council summits, as mentioned above, indicate that the Obama Administration has placed a high priority on the far north. Accordingly, it has reformulated national policy toward the region. As noted elsewhere in this report, the Bush Administration in early January 2009 issued a presidential directive outlining its policy on the Arctic region; the last such directive had been issued in 1994. The Obama Administration operated under the Bush Administration's policy directive until May 10, 2013, when it announced its National Strategy for the Arctic Region.[234] The new policy blueprint identifies three major "lines of effort": (1) Advancing United States security interests, (2) Pursuing responsible Arctic region stewardship, and (3) Strengthening international cooperation. These activities are to be guided by four principles: (1) Safeguard peace and stability, (2) Make decisions using the best available information, (3) Pursue innovative arrangements among various levels of government and the private sector, and (4) Consult and coordinate with Alaska natives.

Security Issues

Throughout the Cold War, the Arctic region was a zone of strategic interest, where the United States, the Soviet Union, and allied states conducted air and naval maneuvers and tested ballistic missiles. With the collapse of the Warsaw Pact and the Soviet Union, however, the importance of the high north diminished in the 1990s. Although the establishment of the Arctic coastal states' sovereignty through the demarcation of boundaries in the region is being conducted peacefully under the auspices of the U.N. Convention on the Law of the Sea, the Arctic is once again being viewed by some as a potential emerging security issue. In a December 2011 *Washington Post* op- ed, Heather Conley, a senior fellow at the Center for Strategic and International Studies, noted several recent developments:

> In April [2011], President Obama signed a new command plan that gives NORAD and the U.S. Northern Command greater responsibility in protecting the North Pole and U.S. Arctic territory.... In 2009, Norway moved its operational command to its northern territories above the Arctic Circle. Russia has plans to establish a brigade that is specially equipped and prepared for military warfare in Arctic conditions. Denmark has made it a strategic priority to form an Arctic Command. Canada is set to revitalize its Arctic fleet, including spending $33 billion to build 28 vessels over the next 30 years.[235]

Canadian academic Rob Huebert has pointed out that in August 2010 the United States, Canada, and Denmark conducted in the Canadian Arctic their annual joint naval exercises involving several advanced and powerful warships. Huebert observed that "while defence officials are quick to point out they see no military threat to the region, it's still interesting to see these three Arctic friends coming together to improve their naval combat capability in the Far North."[236] In varying degrees, the Arctic coastal states have indicated a willingness to establish and maintain a military presence in the high north.[237] Although some have argued that terrorism and hijacking may constitute security concerns in the region, others maintain that such threats are chimerical, given the challenges of distance and geography, and the difficulty of navigating in a polar environment. *The Economist* has asserted that "the risks of Arctic conflict have been exaggerated. Most of the Arctic is clearly assigned to individual countries. According to a Danish estimate, 95% of Arctic mineral resources are within agreed national boundaries."[238] Other factors may also postpone energy exploration. For example, in the *New York Times*, three scholars recently noted that "the shale gas revolution is already delaying some Arctic energy projects." In addition, some companies are reportedly "fearful of the financial and public relations risk of working in the pristine icy wilderness."[239]

The Arctic has also become a region of interest for NATO. However, as one writer has noted, "[t]here is currently no consensus within the alliance that NATO has any role to play in the Arctic, as Canada strongly opposes any NATO involvement on sovereignty grounds and other NATO members are concerned with negative Russian reaction."[240] Speaking in Reykjavík in January 2009, former NATO Secretary General Jaap de Hoop Scheffer urged that member states not allow the Arctic to become a divisive issue. He also recommended that the alliance and Russia cooperate through building upon their shared experience in search-and-rescue operations. Former Danish Prime Minister Anders Fogh Rasmussen, who became secretary general of the alliance in August 2009, has also addressed security in the high north. Citing the "potentially huge security implications" of Arctic climate change, Rasmussen in October 2009 stated that "I think it is within the natural scope of work for NATO to be the forum for consultation and discussion on [selected Arctic] issues." In March 2009, however, Russia's NATO ambassador stated that Moscow would not cooperate with the alliance on Arctic matters. And in September 2010, President Medvedev reportedly observed that "the Arctic can do fine without NATO," and that his government "views [possible NATO] activity with quite serious tension, because it is after all a zone of peaceful cooperation, economic cooperation, and of course the military factor always—at a minimum—creates additional questions." On a visit to Moscow in November 2010, Rasmussen assured the Russians that NATO does not intend to establish a presence in the Arctic.[241] Since 2006, several member and partner states have participated in *Cold Response*, a wide-ranging annual crisis response exercise hosted by Norway. During the most recent joint maneuvers in March 2012, 14 nations fielded air, land, and naval assets and more than 16,000 troops. Although the exercises are multilateral, they are not conducted under the auspices of NATO. According to the Norwegian Defense Ministry's web page, the next exercise will be held in 2014.[242]

A recent NATO Parliamentary Assembly report notes that "50% of the territory surrounding the Arctic Sea is a territory of a NATO member state," and suggests that "NATO could serve as a forum for dialogue on military issues...."[243] The report contends that the alliance is well-equipped to play a key role in addressing security challenges that will likely emerge, particularly those that involve surveillance, search-and-rescue, and environmental cleanup. However, observers note that the lack of unanimity over a NATO presence in the Arctic is reflected by the fact that the high north is mentioned neither in the alliance's 2010 strategic concept, nor in the final declaration of the 2012 Chicago summit. On May 8, 2013, following a visit to Norway, Secretary General Rasmussen stated that "at the present time," the alliance had "no intention of raising its presence and activities in the High North." He later twittered that "the Arctic is a harsh environment. It rewards cooperation, not confrontation. I trust we'll continue to see cooperation."[244]

The Russian government has stated that, although it deplores the notion of an arms race in the region and does not foresee a conflict there, it intends to protect its Arctic interests.[245] However, Russia has at times appeared to be sending out mixed messages in this regard. For example, at the conclusion of a meeting in September 2010, Russian Foreign Minister Sergei Lavrov and Canadian Foreign Minister Lawrence Cannon stated that "any militarization [of the Arctic] is out of the question." And in June 2011, then-Prime Minister (and currently President) Vladimir Putin stated, "Russia will definitely expand its presence in the Arctic. We are open for dialogue with our foreign partners and with all neighbors in the Arctic region. But we will naturally defend our own geopolitical interests firmly and consistently." The following month, Putin announced plans to build a large shipping port on the Yamal peninsula, and the government stated that it would be sending two brigades to the north to protect its interests.

Similarly, some Chinese leaders have voiced concern over perceived emerging security issues in the Arctic. In early March 2010, a Chinese admiral stated that "the current scramble for the sovereignty of the Arctic among some nations has encroached on many other countries' interests," and added that China had to "make short and long term ocean strategic development plans to exploit the Arctic because it will become a future mission for the navy." Some analysts, however, believe that China's general approach toward the Arctic will remain decidedly low-key: "To date, China has adopted a wait-and-see approach to Arctic developments, wary that active overtures would cause alarm in other countries due to China's size and status as a rising global power." China is believed to be keen on resolving through diplomacy the national interests of both littoral and non-Arctic states in the high north. Toward that end, it sought permanent observer status on the Arctic Council. Its candidacy reportedly was delayed by a dispute with Norway, which in 2010 awarded the Nobel Peace Prize to Chinese dissident Liu Xiaobo.[246] However, it was reported in early April 2013 that Norway would support China's bid for observer status at the May Arctic Council meeting. As noted above, China was approved at the meeting for observer status.[247]

A report by the Arctic Institute noted, "The armed forces, beyond their responsibility for handling all contingencies, are also the only agencies with both the requisite monitoring instruments and the physical capabilities to operate in such a vast and inhospitable region."[248] However, as mentioned above, the Arctic Council does not address regional security issues. To fill this apparent void, a report by the Washington D.C.- based Center for Strategic and International Studies has proposed the creation of a separate organization, the Arctic Coast Guard Forum (ACGF), consisting initially of the eight Arctic Council states, but possibly expanding eventually to include other countries willing to contribute assets. The ACGF, which could potentially be headquartered at the U.S. Air Force base in Thule, Greenland, would "focus first on information sharing yet should also seek to develop methods of cooperation in support of the Arctic Council's search-and-rescue agreement and future international oil spill response agreement."[249]

However, other, relatively little-publicized multilateral discussions of security issues have already been taking place. In mid-2011, the U.S. European Command (EUCOM), in cooperation with the Norwegian Ministry of Defense, established the Arctic Security Forces Roundtable (ASFR), consisting of high-ranking military officers from the eight members of the Arctic Council, plus France, Germany, the Netherlands, and the UK. Their first meeting, held in June 2011 in Oslo, addressed a range of issues, including infrastructure, the environment, joint exercises and training, and marine domain awareness. In August 2012, the ASFR held a conference Bodo, Norway; the meeting focused mainly on how to improve the communications infrastructure in the high north.[250] Another newly formed venue at which military leaders discuss Arctic issues is the Northern Chiefs of Defense conference, the first of which was held in Goose Bay, Labrador, in May 2012; it was attended by military representatives from the eight Arctic Council governments.[251]

U.S. Military Forces and Operations[252]

During the Cold War, the Arctic was an arena of military competition between the United States and the Soviet Union, with both countries, for example, operating nuclear-powered submarines, long-range bombers, and tactical aircraft in the region. The end of the Cold War and the collapse of most elements of the Russian military establishment following the dissolution of the Soviet Union in December 1991 greatly reduced this competition and led to a reduced emphasis on the Arctic in U.S. military planning.

The diminishment of Arctic sea ice is now leading U.S. military forces to pay renewed attention to the Arctic. This is particularly true in the case of the Navy and Coast Guard, for whom diminishment of Arctic sea ice is opening up potential new operating areas for their surface ships. Navy and Coast Guard activities relating to the Arctic are taking place as other countries, such as Canada, Russia, and Norway, are examining the potential implications for their military forces of diminished Arctic sea ice, and taking or contemplating steps to increase their own navy and coast guard presence and operations in the region. Defense officials in the United States and other countries view issues such as sovereignty, freedom of navigation, and energy exploration as creating a potential in the Arctic for military cooperation, competition, or conflict, depending on how these issues are handled.[253]

DOD in General

2010 Quadrennial Defense Review

The Department of Defense's (DOD's) report on the 2010 Quadrennial Defense Review (QDR), submitted to Congress in February 2010, states:

> The effect of changing climate on the Department's operating environment is evident in the maritime commons of the Arctic. The opening of the Arctic waters in the decades ahead[,] which will permit seasonal commerce and transit[,] presents a unique opportunity to work collaboratively in multilateral forums to promote a balanced approach to improving human and environmental security in the region. In that effort, DoD must work with the Coast Guard and the Department of Homeland

Security to address gaps in Arctic communications, domain awareness, search and rescue, and environmental observation and forecasting capabilities to support both current and future planning and operations. To support cooperative engagement in the Arctic, DoD strongly supports accession to the United Nations Convention on the Law of the Sea.[254]

The report also states:

The Department of Defense and its interagency partners must be able to more comprehensively monitor the air, land, maritime, space, and cyber domains for potential direct threats to the United States. Such monitoring provides the U.S. homeland with an extended, layered in depth defense. This effort includes enhanced coordination with Canada for the defense of North America as well as assisting Mexico and Caribbean partners in developing air and maritime domain awareness capacities. Special attention is required to develop domain awareness tools for the Arctic approaches as well. In coordination with domestic and international partners, DoD will explore technologies that have the potential to detect, track, and identify threats in these spheres to ensure that capabilities can be deployed to counter them in a timely fashion.[255]

The report further states:

Central to the security of the United States is a strong transatlantic partnership, which is underpinned by the bilateral relationships between the United States and the governments of Europe. We will continue to work with this community of like-minded nations, whether by engaging with allies still shaping their democracies after decades of living in the shadow of the Soviet Union, building on the benefits of French reintegration into NATO's military structure, or addressing new security issues such as those arising in the Arctic region.[256]

The report states: "We will seek out opportunities to work with Moscow on emerging issues, such as the future of the Arctic" and that DOD "will also enhance defense relationships and continue to work with Canada in the context of regional security, increased interaction in the Arctic, and combat operations in Afghanistan."[257]

April 2011 Change to Unified Command Plan

In April 2011, it was reported that

> Changes made to the U.S. military's Unified Command Plan shift geographic boundaries and stress the growing importance of the Arctic, officials said.
>
> President Barack Obama signed the document yesterday [April 6].
>
> The biggest change to the plan assigns U.S. Northern Command responsibility for the Arctic. U.S. European Command and U.S. Pacific Command shared responsibility with U.S. Northern Command for the region under the last change published in December 2008. It also places responsibility for Alaska under Northern Command. The previous plan had Northern Command and U.S. Pacific Command sharing responsibility for the state and adjacent waters.
>
> "Northcom was given advocacy responsibility for Arctic capabilities primarily due to having the only U.S. Arctic territory within its area of operations," a Pentagon spokesman said.
>
> Northern Command also already works closely with Canada and "has a habitual relationship with the Department of Homeland Security and the U.S. Coast Guard," the spokesman continued. "These relationships are key to human and environmental safety and security."...
>
> The Unified Command Plan is the responsibility of the chairman of the Joint Chiefs of Staff, and is reviewed every two years. The Joint Staff coordinates input from the combatant commanders, the service chiefs and Defense Department leadership. The chairman, Navy Adm. Mike Mullen, submitted his recommendations through Defense Secretary Robert M. Gates to Obama.[258]

May 2011 Report to Congress

A May 2011 DOD report to Congress on Arctic operations and the Northwest Passage that was prepared at congressional direction[259] stated:

> The Arctic is warming on average twice as fast as the rest of the planet, resulting in increased human activity in the region. Although some perceive that competition for resources and boundary disputes may result in conflict in the Arctic, the opening of the Arctic also presents opportunities to work collaboratively in multilateral forums to promote a balanced approach to improving human and environmental security in the region.
>
> Strategic guidance on the Arctic is articulated in National Security Presidential Directive (NSPD) 66 / Homeland Security Presidential Directive (HSPD) 25, *Arctic Region Policy*. Additional guidance is found in *the 2010 National Security Strategy (NSS)* and the *2010 Quadrennial Defense Review (QDR)*. The overarching strategic national security objective is *a stable and secure region where U.S. national interests are safeguarded and the U.S. homeland is protected.* This objective is consistent with a regional policy that reflects the relatively low level of

threat in a region bounded by nation states that have not only publicly committed to working within a common framework of international law and diplomatic engagement, but also demonstrated ability and commitment to doing so over the last fifty years.

DoD will take responsible steps to anticipate and prepare for the Arctic operations of the near-(2010-2020), mid- (2020-2030), and far-term (beyond 2030). Capabilities will need to be reevaluated as conditions change, and gaps must be addressed in order to be prepared to operate in a more accessible Arctic. Key challenges include: shortfalls in ice and weather reporting and forecasting; limitations in command, control, communications, computers, intelligence, surveillance, and reconnaissance (C4ISR) due to lack of assets and harsh environmental conditions; limited inventory of ice-capable vessels; and limited shore-based infrastructure. The key will be to address needs in step with the rate at which activity in the Arctic increases, and balance potential investments in these capabilities with other national priorities. The United States has a vital Arctic neighbor and partner in Canada, with its shared values and interests in the region. DoD will work with the Canadian Department of National Defence (DND) to ensure common Arctic interests are addressed in a complementary manner

Although having multiple Combatant Commanders (CCDRs) in the Arctic Ocean makes coordination more challenging, having too few would leave out key stakeholders, diminish longstanding relationships, and potentially alienate important partners. There are now two CCDRs with Arctic responsibilities: Commander, U.S. European Command (USEUCOM), and Commander, U.S. Northern Command (USNORTHCOM), each responsible for a portion of the Arctic Ocean aligned with adjacent land boundaries, an arrangement judged best suited to achieve continuity of effort with key regional partners.

Existing DoD posture in the region is adequate to meet near- to mid-term U.S. defense needs. DoD does not currently anticipate a need for the construction of a deep-draft port in Alaska between now and 2020. Given the long lead times for construction of major infrastructure in the region, DoD will periodically re-evaluate this assessment as the Combatant Commanders update their regional plans on a regular basis.

The United States needs assured Arctic access to support national interests in the Arctic. This access can be provided by a variety of proven capabilities, including submarines and aircraft, but only U.S.-flagged ice-capable ships provide visible U.S. sovereign maritime presence throughout the Arctic region. This need could potentially be met by either icebreakers or ice- strengthened surface vessels, none of which are in the U.S. Navy current surface combatant inventory, but which do exist in U.S. Coast Guard's inventory in limited numbers.

Finally, significant uncertainty remains about the rate and extent of climate change in the Arctic and the pace at which human activity will

> increase. The challenge is to balance the risk of being late-to-need with the opportunity cost of making premature Arctic investments. Not only does early investment take resources from other pressing needs, but the capabilities would be later in their lifecycle when finally employed. Given the many competing demands on DoD's resources in the current fiscal environment, the Department believes that further evaluation of the future operating environment is required before entertaining significant investments in infrastructure or capabilities.[260]

January 2012 GAO Report Reviewing May 2011 DOD Report

A January 2012, congressionally directed[261] GAO report reviewing the May 2011 DOD report above stated the following:

> While DOD has undertaken some efforts to assess the capabilities needed to meet national security objectives in the Arctic, it is unclear whether DOD will be in a position to provide needed capabilities in a timely and efficient manner because it lacks a risk-based investment strategy for addressing near-term needs and a collaborative forum with the Coast Guard for addressing long-term capability needs. DOD's [May 2011] Arctic Report acknowledges that it has some near-term gaps in key capabilities needed to communicate, navigate, and maintain awareness of activity in the region. However, DOD has not yet evaluated, selected, or implemented alternatives for prioritizing and addressing near-term Arctic capability needs. In addition, DOD and the Coast Guard have established a working group to identify potential collaborative efforts to enhance U.S. Arctic capabilities. This working group is focused on identifying potential near-term investments but not longer-term needs, and it is currently expected to be dissolved in January 2012. Uncertainty involving the rate of Arctic climate change necessitates careful planning to ensure efficient use of resources in developing Arctic needs such as basing infrastructure and icebreakers, which require long lead times to develop and are expensive to build and maintain. Without taking steps to meet near- and long-term Arctic capability needs, DOD risks making premature Arctic investments, being late in obtaining needed capabilities, or missing opportunities to minimize costs by collaborating on investments with the Coast Guard....
>
> DOD has several efforts under way to assess the capabilities needed to support U.S. strategic objectives in the Arctic. However, it has not yet developed a comprehensive approach to addressing Arctic capabilities that would include steps such as developing a risk-based investment strategy and timeline to address near-term needs and establishing a collaborative forum with the Coast Guard to identify long-term Arctic investments....

According to DOD's Arctic Report, capabilities will need to be reassessed as conditions change, and gaps will need to be addressed to be prepared to operate in a more accessible Arctic. Other stakeholders have also assessed Arctic capability gaps. Examples of these efforts include the following:

- U.S. Northern Command initiated a commander's estimate for the Arctic in December 2010 that, according to officials, will establish the commander's intent and missions in the Arctic and identify capability shortfalls. In addition, Northern Command identified two Arctic-specific capability gaps (communications and maritime domain awareness) in its fiscal years 2013 through 2017 integrated priority list, which defines the combatant command's highest-priority capability gaps for the near-term, including shortfalls that may adversely affect missions.
- U.S. European Command completed an Arctic Strategic Assessment in April 2011 that, among other things, identified Arctic capability gaps in the areas of environmental protection, maritime domain awareness, cooperative development of environmental awareness technology, sharing of environmental data, and lessons learned on infrastructure development. In addition, it recommended that the command conduct a more detailed mission analysis for potential Arctic missions, complete a detailed capability estimate for Arctic operations, and work in conjunction with Northern Command and the Departments of the Navy and Air Force to conduct a comprehensive capabilities-based assessment for the Arctic.
- DOD and DHS established the Capabilities Assessment Working Group (working group) in May 2011 to identify shared Arctic capability gaps as well as opportunities and approaches to overcome them, to include making recommendations for near-term investments. The working group was directed by its Terms of Reference to focus on four primary capability areas when identifying potential collaborative efforts to enhance Arctic capabilities, including near-term investments. Those capability areas include maritime domain awareness, communications, infrastructure, and presence. The working group was also directed to identify overlaps and redundancies in established and emerging DOD and DHS Arctic requirements. As the advocate for Arctic capabilities, Northern Command was assigned lead responsibility for DOD in the working group, while the Coast Guard was assigned lead responsibility for DHS. The establishment of the working group—which, among other things, is to identify opportunities for bi-departmental action to close Arctic capability gaps and issue recommendations for near-term investments—helps to ensure that collaboration between the Coast

Guard and DOD is taking place to identify near-term capabilities needed to support current planning and operations. Although the working group is developing a paper with its recommendations, officials indicated that additional assessments would be required to address those recommendations.

- U.S. Navy completed its first Arctic capabilities-based assessment in September 2011 and is developing a second capabilities-based assessment focused on observing, mapping, and environmental prediction capabilities in the Arctic, which officials expect to be completed in the spring of 2012. The Navy's first Arctic capabilities-based assessment identified three critical capability gaps as the highest priorities, including the capabilities to provide environmental information; maneuver safely on the sea surface; and conduct training, exercise, and education. This assessment recommended several near-term actions to address these gaps....

Even though DOD has made preliminary efforts to identify Arctic capability gaps and assess strategic objectives, constraints, and risks in the Arctic, DOD has not yet evaluated, selected, or implemented alternatives for prioritizing and addressing near-term Arctic capability needs....

Given that the opening in the Arctic presents a wide range of challenges for DOD, a risk- based investment strategy and timeline can help DOD develop the capabilities needed to meet national security interests in the region. Without a risk-based investment strategy and timeline for prioritizing and addressing near-term Arctic capability gaps and challenges, which is periodically updated to reflect evolving needs, DOD could be slow to develop needed capabilities, potentially facing operational risk and higher costs if the need arises to execute plans rapidly. Conversely, DOD could move too early, making premature Arctic investments that take resources from other, more pressing needs or producing capabilities that could be outdated before they are used.

While DOD and DHS have established the working group to identify shared near-term Arctic capability gaps, this collaborative forum is not intended to address long-term Arctic capability gaps or identify opportunities for joint investments over the longer-term. DOD acknowledged the importance of collaboration with the Coast Guard over the long-term in its 2010 Quadrennial Defense Review, which states that the department must work with the Coast Guard and DHS to develop Arctic capabilities to support both current and future planning and operations. According to DOD and Coast Guard officials, although the working group is primarily focused on near-term investments, it has discussed some mid- to long- term capability needs. However, DOD and Coast Guard officials stated that after the completion of the working group's paper, expected in January 2012, the working group will have completed the tasks detailed in the Terms of Reference and will be dissolved. Consequently, no forum will exist to further address any mid- to long-term capability needs....

After the working group completes its tasks in January 2012, there will be no DOD and Coast Guard organization focused specifically on reducing overlap and redundancies or collaborating to address Arctic capability gaps in support of future planning and operations, as is directed by the 2010 Quadrennial Defense Review.

While Northern Command officials stated they have plans for periodic reassessment of long- term capability needs, such as icebreakers or basing infrastructure including a deep-water port, it is not clear how those plans consider collaboration with the Coast Guard. For example, officials stated the biennial review of Northern Command's Theater Campaign Plan and Strategic Infrastructure Master Plan will consider long-term capability and infrastructure needs. They added that the commander's Arctic Estimate is reviewed annually and also considers long-term priorities, such as identifying a need for icebreakers. However, the officials stated that the Arctic Estimate does not identify how DOD would acquire those icebreakers or how it would coordinate with the Coast Guard—the operator of the nation's icebreakers—to reconstruct existing or build new icebreakers....

Without specific plans for a collaborative forum between DOD and the Coast Guard to address long-term Arctic capability gaps and to identify opportunities for joint investments over the longer-term, DOD may miss opportunities to leverage resources with the Coast Guard to enhance future Arctic capabilities.

At this time, significant DOD investments in Arctic capabilities may not be needed, but that does not preclude taking steps to anticipate and prepare for Arctic operations in the future. Addressing near-term gaps is essential for DOD to have the key enabling capabilities it needs to communicate, navigate, and maintain awareness of activity in the region. An investment strategy that identifies and prioritizes near-term Arctic capability needs and identifies a timeline to address them would be useful for decision makers in planning and budgeting. Without taking deliberate steps to analyze risks in the Arctic and prioritize related resource and operational requirements, DOD could later find itself faced with urgent needs, resulting in higher costs that could have been avoided.

In addition, unless DOD and DHS continue to collaborate to identify opportunities for interagency action to close Arctic capability gaps, DOD could miss out on opportunities to work with the Coast Guard to leverage resources for shared needs. DOD may choose to create a new collaborative forum or incorporate this collaboration into an existing forum or process. Given the different missions and associated timelines of DOD and the Coast Guard for developing Arctic capabilities, it is important that the two agencies work together to avoid fragmented efforts and reduce unaffordable overlap and redundancies while addressing Arctic capability gaps in support of future planning and operations.

To more effectively leverage federal investments in Arctic capabilities in a resource- constrained environment and ensure needed capabilities are developed in a timely way, we recommend that the Secretary of Defense, in consultation with the Secretary of the Department of Homeland Security, take the following two actions:

- develop a risk-based investment strategy that: 1) identifies and prioritizes near-term Arctic capability needs, 2) develops a timeline for addressing them, and 3) is updated as appropriate; and
- establish a collaborative forum with the Coast Guard to fully leverage federal investments and help avoid overlap and redundancies in addressing long-term Arctic capability needs.

In written comments on a draft of this report, DHS concurred with both of our recommendations. For its part, DOD partially concurred with both of our recommendations. It generally agreed that the department needed to take action to address the issues we raised but indicated it is already taking initial steps to address them.[262]

June 2012 Senate Armed Services Committee Report Language

The Senate Armed Services Committee, in its report (S.Rept. 112-173 of June 4, 2012) on S. 3254, the FY2013 National Defense Authorization Act, stated:

Arctic region

The committee recognizes the continued importance of the Arctic region to our broader national strategy. Declining ice cover continues to open the Arctic region and a concerted, systematic, and immediate effort should be undertaken to adequately protect the United States' security, environmental, energy, economic, and natural resource interests in the Arctic.

The committee commends the Department of Defense (DOD) and, in particular, U.S. Northern Command and the Department of the Navy, and other federal agencies, such as the Department of Homeland Security (DHS), through the U.S. Coast Guard, the Department of Commerce, through the National Oceanic and Atmospheric Administration and the U.S. Arctic Research Commission, for their progress thus far in studying the region and defining the capabilities required to effectively operate in and protect the domain.

International and interagency collaboration is also needed to develop the necessary mapping and charting resources required for safe navigation and to promote security and economic interests. As sea ice recedes, timely weather forecasts and disaster warnings along with more baseline data will be required to conduct successful search and rescue missions. Search and rescue coordination, planning, and training for the Arctic should be thoroughly analyzed and developed to ensure forces can successfully operate in the domain.

The committee recognizes the importance of DOD's involvement in interagency and international efforts to protect national security interests in the region and, accordingly, urges DOD to continue to work in concert with DHS to establish a formal chartered working group to pursue increasing Arctic capabilities in the areas of communications, maritime domain awareness, infrastructure, and presence, as was recommended by U.S. Northern Command and the U.S. Coast Guard in a joint White Paper dated March 13, 2012. The committee also urges the agencies to develop an investment strategy for funding emerging requirements in balance with a resource constrained environment. (Pages 194-195)

August 2013 Senate Appropriations Committee Report Language

The Senate Appropriations Committee, in its report (S.Rept. 113-85 of August 1, 2013) on S. 1429, the FY2014 DOD Appropriations Act, stated:

ENGAGEMENT ON ARCTIC ISSUES

The Committee notes that the physical changes in the Arctic are unprecedented in both their rate and scope of change. In addition to the economic and social concerns, numerous studies, including the 2010 Quadrennial Defense Review and the U.S. Navy Arctic Roadmap, have documented the significant impact that a rapidly changing Arctic has on National Security, to include an increasing number of vessels expected to be operating in the Arctic Ocean in the summer and fall months.

The Arctic Council has grown significantly in recent years with increased influence, visibility, and membership. As of May 2013, China, Japan, Singapore, India, The Republic of Korea and Italy have been granted observer status. In 2015, the U.S. will assume its 2-year Chair of the Arctic Council, succeeding Canada in this role. In view of the increasing significance and importance of the Arctic Council, and the United States' upcoming leadership in this body, the Committee urges the Secretary of Defense to continue to examine ways the Department can support engagement on Arctic issues, including funding for better Arctic Domain Awareness, mapping the U.S. Arctic waters, and enhanced observations and prediction of Arctic weather, ocean, and ice conditions. (Page 14)

Navy and Coast Guard

The Navy and Coast Guard are exploring the potential implications that increased surface ship and aircraft operations in the Arctic may have for required numbers of ships and aircraft, ship and aircraft characteristics, new or enlarged Arctic bases, and supporting systems, such as navigation and communication systems. The Navy and Coast Guard have sponsored or participated in studies and conferences to explore these implications, the Coast Guard has deployed cutters and aircraft into the region to perform missions and better understand the implications of operating such units there, and Navy sailors have ridden on Canadian Navy ships deploying to Arctic waters for similar reasons.

Points or themes that have emerged in studies, conferences, and deployments regarding the potential implications for the U.S. Navy and Coast Guard of diminished Arctic sea ice include but are not limited to the following:

- The diminishment of Arctic ice is creating potential new operating areas in the Arctic for Navy surface ships and Coast Guard cutters.
- U.S. national security interests in the Arctic include "such matters as missile defense and early warning; deployment of sea and air systems for strategic sealift, strategic deterrence, maritime presence, and maritime security operations; and ensuring freedom of navigation and overflight."[263]
- Search and rescue in the Arctic is a mission of increasing importance, particularly for the Coast Guard, and one that poses potentially significant operational challenges (see "Search and Rescue" above).
- More complete and detailed information on the Arctic is needed to more properly support expanded Navy and Coast Guard ship and aircraft operations in the Arctic.
- The Navy and the Coast Guard currently have limited infrastructure in place in the Arctic to support expanded ship and aircraft operations in the Arctic.
- Expanded ship and aircraft operations in the Arctic may require altering ship and aircraft designs and operating methods.
- Cooperation with other Arctic countries will be valuable in achieving defense and homeland security goals.

Navy in General

In a February 2009 journal article, the Oceanographer of the Navy stated:

> Preserving freedom of navigation in the region is an important tenet of U.S. policy. The NSR [Northern Sea Route], however, is a contested

waterway, with Russian claims of sovereignty competing against U.S. and European Union insistence that it is an international strait available to all nations, subject to mutually recognized terms. Another potential [Arctic] transoceanic shipping route may be the Northwest Passage, which extends from the Atlantic through Baffin Bay and the Canadian Archipelago and into the Pacific by way of the Bering Strait. Canada claims sovereignty over the waters of the Canadian Archipelago, although the United States and the European Union claim that the Northwest Passage also constitutes an international strait which allows right of innocent passage....

Aside from access and right of passage, the Navy and Coast Guard, in particular, must also be concerned with strategic choke points such as the Bering Strait, Canada's Queen Elizabeth Islands in the Northwest Passage, and Russia's Severnaya Zemlya and New Siberian Islands in the Northern Sea Route. These narrow passages offer some protection from persistent ice blockage, but they are also vulnerable to control or blockade by adversaries that would significantly disrupt potential commercial shipping and oil transport....

U.S. naval interests will face new challenges in an increasingly ice-free Arctic with a strategic objective to understand potential threats to the United States from the maritime domain. As throughout the global commons, the U.S. Navy must be aware of activities that could be harmful to national security interests in a region that will, no doubt, see fewer barriers to access by potential adversaries in the future. National and homeland security interests pertinent to the U.S. Navy in the region would include early warning/missile defense; maritime presence and security; and freedom of navigation and over-flight....

The region is primarily a maritime domain and the U.S. Navy of the future must be prepared to protect sea lines of communication supporting maritime commerce and other national interests—including national security—there. In addition to thinking through how we adjust our shipbuilding emphasis to support such operations, the Navy should also be thinking strategically about building the necessary infrastructure to provide logistic support for Arctic patrols, search and rescue capabilities, and shore-based support activities.

To ensure complete maritime domain awareness in the region, and to provide our forces a competitive advantage, it will be necessary to have comprehensive knowledge of the physical environment. Data must be obtained by a suite of remote sensors (satellites, radars), autonomous sensors (data buoys, unmanned vehicles), and manned sensors (shipboard, coastal observing stations). Computer-based ocean and atmospheric models must be adjusted to the geophysical peculiarities of high latitudes. Communication lines for data exchange and reach-back processing at high-performance computing production centers must be robust and reliable. To ensure safety of navigation, we will also need to conduct more high-resolution bottom surveys and increase the scrutiny we place on sea ice conditions.

The Navy relies on its international and interagency partners for assistance to ensure success of maritime domain awareness and maritime security missions. To meet the demands of national security in the changing northern environment, strengthening mechanisms for cooperation among the regional nations and U.S. agencies must remain a high priority. Like everywhere else in the world, sound national security in the Arctic will require strategic access, military mobility, safe navigation, unimpeded maritime transportation, improved homeland security, and responsible, sustainable use of ocean and coastal resources. International and interagency agreements and partnerships are vital to incorporating these essential elements into a viable national security policy and will be critical for resolving future naval challenges of a changing Arctic.[264]

In a spring 2010 journal article, the Oceanographer of the Navy (the successor to the officer who authored the above journal article) and a coauthor stated that "Navy surface vessels are able to operate up to the marginal ice zone but will require ice-strengthening to operate in higher ice conditions; Navy aircraft are capable of operating in the Arctic, but the lack of divert fields limits their duration and range."[265] The article also stated:

While the Navy has a rich history in the Arctic, several challenges must be met to ensure successful operations in the future. These include the lack of support infrastructure and logistics support, environmental hazards such as drifting sea ice and icing on exposed surfaces, and communications difficulties. Antiquated nautical charts, drifting ice, low visibility, and the paucity of electronic and visual navigation aids hinder safety of navigation. A lack of coastal installations also contributes to the difficulty of search and rescue (SAR) operations. The only American-owned deepwater port near the Arctic basin is Dutch Harbor, in the Aleutian Islands.[266]

On June 16, 2011, in remarks at a seminar on the Arctic, Admiral Gary Roughead, the Chief of Naval Operations, stated:

Well in my mind, there is a phenomenal event taking place on the planet today, and that is what I call the opening of the Fifth Ocean; that's the Arctic Ocean. We haven't had an ocean open on this planet since the end of the Ice Age. So if this is not a significant change that requires new, and I would submit, brave thinking on the topic, I don't know what other sort of physical event could produce that.

> We in the Navy a couple of years ago began to look at how we as a Navy serve the nation.... So as we began to look at the planet, we realized that there were significant changes taking place; opening of the Fifth Ocean; physical changes; changes in population as they compressed into the littoral areas.
>
> And so we established, I established something that I call Task Force Climate Change. Primarily focused on the Arctic because of the massive changes that are taking place up there, but it also takes into account what is happening on the rest of the planet and where we as a Navy think we have to be in the future. And then from that Task Force Climate Change, we developed what we call the Arctic road map....
>
> The way we see things taking place is the first of moves, which will largely be in the area of resources. Fishing stocks will move with the water temperature, they will start to drift farther north, that will take fishing fleets farther north and there is a set of responsibilities that we have with other nations on making sure that that activity is taking place lawfully.
>
> Then we believe the next step will be the extraction of resources. So what do we have to be prepared to do as a Navy, as a military particularly aligned closely with the Coast Guard on this.
>
> And then in about 20 years time, 25 years time, the Arctic becomes a profitable sea route from Asia to Europe over the top of the planet. If you look at some of the estimates from shippers, some may actually be here.
>
> That is about a million dollars a trip someone saves. That is not insignificant. And so it will become a busy place as transportation starts to take place. That is how we envision the changes that are taking place, how we are looking at them, and the road map that we put in place is based on the best science as we know today. And we are committed to providing the resources the nation needs from the naval perspective to meet our needs.[267]

Four days later, in remarks to a symposium on the impacts on naval and maritime operations of an ice-diminishing Arctic, Admiral Roughead stated:

> Since this forum began in 2001, our Navy has benefitted from the development and the discussion of the Arctic Symposium, as it has improved our appreciation greatly for an ice- diminished Arctic and the security implications that will follow from that physical change that will take place.
>
> This broader examination is something that we've endeavored to focus on systematically since we stood up our Task Force Climate Change in May 2009, and issued our Arctic Roadmap that same year. And rather than all the answers, what our efforts have given us is an

appreciation of just how dynamic the study of the region is and surely will be in the future....

... at a fundamental level, the trends point undeniably towards a new venue of operations and responsibility for our global Navy—for preserving American interests in free and fair access there—and in light of this we remain committed to preparing exhaustively for the challenges and especially for the opportunities that are going to exist in an ice-diminishing Arctic.

The U.S. Navy's interests in the Arctic are not new, of course. We have many decades of experience with exploration and, indeed, episodic operations in the waters of the Arctic Circle....

But never has our interest encountered the confluence of trends, as projected by the U.S. Geological Survey in 2008 and the National Research Council this past March, that promises to change the Arctic so pervasively, and in so doing affect the global environment for which we plan and program our future fleet....

In projecting the impact of climate change in the northern latitudes, however, I'm reminded of what [then-NOAA Administrator] Dr. [Jane] Lubchenco observed just this past March, when she said, "what happens in the Arctic does not stay there." The trends we discuss here, in a similar timeframe, promise more disruption and disorder in a world whose population is growing rapidly, and moving to megacities on or near the coasts of almost every continent. The prospects of sea level rise, for some megacities, or the coral islands of the Maldives, are similarly daunting. We also have to consider the likely frictions that arise as fishing stocks migrate with changing sea temperatures, and the very real possibility that conflicts in the future will be fought over access to dwindling natural supplies of fresh water.

It is because of these projections that our Navy is preparing for increased demand, both in the region—where we will maintain our access and uphold the freedom of navigation as a global good—and beyond, where we expect developments to expose the costs of our national reluctance on the Law of the Sea convention and to test our present understanding of customary legal guarantees to the very freedoms behind our global operations today. We are considering the technical requirements for polar operations to support our strategic objective of a safe, stable, and secure Arctic region where our national interests are safeguarded— namely, how and when to build forces capable and competent for the harsh northern climes.

We also remain well aware of how important maritime partnerships will be in addressing capacity concerns as we seek to reinforce universal values in the global commons. Some speak of a changing Arctic in terms of either a race or a zero-sum game. As far as the U.S. Navy is concerned, it is neither. Rather, we see it as an opportunity to extend to the 'Fifth Ocean' the principles that have benefitted all peaceful nations in the other four. Secretary of State Clinton's attendance at the meeting of Arctic Council ministers last month in Nuuk, Greenland, signals the United States6 level of interest in the region, as well as our belief in the power of international cooperation to advance mutual interests in the free use of the maritime domain. It was also inherently practical in that it forged the much-needed agreement among Arctic states to improve our coordination on search and rescue.

As we stated in our Arctic Roadmap, the Navy will continue to pursue cooperative relationships around such areas of common interest, whether mission enablers like Maritime Domain Awareness, core naval capabilities such as humanitarian assistance and disaster response, or intra-governmental relationships like defense support to civil authorities. Exercises such as the Canadian Navy's 'Operation Nanook,' which we observed in 2009 and had the privilege of participating in last August alongside our Coast Guard counterparts, enhance our preparedness in several of these areas, and lay the basis from which we might extend the same maritime partnership and interoperability we have come to value so highly in our global operations. Joint exercises such as the U.S. Pacific Command's 'Northern Edge'—which started just outside of the Arctic Circle one week ago today and will continue throughout this week – ensure that we as a Navy remain ready to partner at the high end of operations as required.[268]

An April 28, 2012, press report states:

As global warming opens the Arctic Ocean to commercial and industrial traffic, the U.S. Navy is pushing to catch up with Russia, Canada and even Denmark in its Arctic ability. If a crisis were to happen now, the Navy lacks the ability to act in the Arctic without the help of one of those countries or the Coast Guard.

Last year, the Navy asked the War Gaming Department of the U.S. Naval War College to find out what the Navy needs for sustained operations in the Arctic.

In the resulting 2011 Fleet Arctic Operations Game, the Navy learned how big its Arctic shortcomings are. As a force, the Navy lacks everything from bases and Arctic-capable ships to reliable communications and cold-weather clothing....

The game's conclusions: the Navy is not adequately prepared to conduct long-term maritime Arctic operations; Arctic weather conditions increase the risk of failure; and most critically, to operate in the Arctic, the Navy will need to lean on the U.S. Coast Guard, countries like Russia or Canada, or tribal and industrial partners.

To sustain operations in the Arctic, the Navy needs ice-capable equipment, accurate and timely environmental data, personnel trained to operate in extreme weather, and better communications systems. Much of the environmental data will come from other Arctic nations....

Navy officials understand the need to conduct exercises in the Arctic so they can get ready for the real thing, but they don't have a strategy.

"We are the only Arctic nation without an Arctic strategy," said U.S. Navy Cmdr. Blake McBride, Arctic Affairs officer for Task Force Climate Change. "The Coast Guard and Department of Defense are working on a strategy to help answer the issue, and advocate for capabilities."

Aside from signing National Security Presidential Directive 66, which requires the U.S. to have a presence in the Arctic, the Arctic hasn't been a priority for the U.S. government, largely because there isn't an immediate military threat.

"It's becoming a higher priority, but we don't make our own priorities," McBride said. "We don't foresee a military threat in the Arctic, but it doesn't mean we will not need to be able to operate there."

The Navy's future plans to conduct operations in the Arctic largely depend on the budget.

"It's all about the money," McBride said. "If you don't have the budget or funds to invest in manpower and equipment then you don't have anything."[269]

2009 Navy Arctic Roadmap

The Navy examined issues and concerns relating to climate change at a May 15, 2009, meeting of the Chief of Naval Operations (CNO) Executive Board. Following this meeting, the Navy decided to establish a Navy group led by the Oceanographer of the Navy called Task Force Climate Change (TFCC), and to develop Navy roadmaps first for the Arctic, and later for more general responses to global climate change.

The Navy issued its Arctic roadmap on November 10, 2009.[270] The document, which is dated October 2009 and co-sponsored by TFCC and the Oceanographer of the Navy,[271] is intended to guide the service's activities regarding the Arctic for the period FY2010-FY2014. The November 10, 2009, cover memo states that the document was to remain in effect until the completion of the report on the FY2014 Quadrennial Defense Review (QDR), at which time it would be reviewed and revised to incorporate guidance from that QDR.

2013 Update to Navy Arctic Roadmap

In his prepared statement for an April 16, 2013, hearing before the House Armed Services Committee, Admiral Jonathan Greenert, the Chief of Naval Operations, stated:

Emerging projections assess that the Arctic will become passable for shipping several months out of the year within the next decade—about 10 years earlier than predicted in 2009 when we first published our Arctic Roadmap. This will place new demands on our fleet for presence in the Arctic and capabilities to operate in the Arctic environment. Between now and the start of FY2014 we will update our Arctic Roadmap, and accelerate many of the actions Navy will take in preparation for a more accessible Arctic. During FY2014 we will implement this revised roadmap, including developing with the U.S. Coast Guard plans for maintaining presence and search and rescue capability in the Arctic and pursuing exchanges with other Arctic countries to familiarize our Sailors with Arctic operations.[272]

August 2011 Navy Arctic Environmental Assessment and Outlook Report

In August 2011, the Navy released an Arctic environment assessment and outlook report.[273] The report states that it

addresses Action Item 5.7 of the U.S. Navy Arctic Roadmap - to produce an Arctic Environmental Assessment and Outlook Report. This first biennial report provides a comprehensive assessment of the state of the Arctic environment, including the oceanography, hydrography, meteorology, fisheries, ice-extent, and climatic trends. This is important because the IPCC [reporting] refresh rate is too long to meet the budget POM [Program Objective Memorandum] cycle,274 so this assessment will periodically synthesize existing scientific reports to inform POMs, specifically POM-[FY20]14; this allows the Navy's decisions to be based on sound science, and not use one source only, but a consensus of accepted sources....

As the Arctic environment continues to change and human activity increases, the U.S. Navy must be prepared to operate in this region. It is important to note that even though the Arctic is opening up, it will continue to be a harsh and challenging environment for the foreseeable future due to hazardous sea ice, freezing temperatures and extreme weather. Although the Navy submarine fleet has decades of experience operating in the Arctic, the surface fleet, air assets, and U.S. Marine Corps ground troops have limited experience there. The Navy must now consider the Arctic in terms of future policy, strategy, force structure, and investments.[275]

An April 28, 2012, press report states:

Navy officials have done the work called for in phase one and two of the [Arctic] roadmap, which largely consisted of developing research,

> assessing fleet readiness, completing capabilities-based assessments like the Fleet Arctic Operations Game, and formalizing cooperative agreements.
>
> The biggest hurdle comes in the next phase, which calls for funding equipment and Arctic training. Navy officials say they are drafting a budget request to address those items.[276]

June 2013 House Armed Services Committee Report Language

The House Armed Services Committee, in its report (H.Rept. 113-102 of June 7, 2013) on H.R. 1960, the FY2014 National Defense Authorization Act, stated:

> Navy Arctic Roadmap
>
> The committee continues to be concerned about the Department of Defense's resources and preparedness for accessing, operating in, and protecting national interests in the Arctic. The Navy currently estimates that between 2020 and 2030, the Arctic could be ice free for one month during the summer which may lead to an increase in trans-Arctic passage for vessels seeking to reduce transit distance by utilizing the Northern Sea Route and the Northwest Passage. The Navy's Strategic Objectives for the Arctic was signed in May 2010 and is referred to in both the Navy Arctic Capabilities Based Assessment, approved in September 2011, and the Navy Arctic Environmental Capabilities Based Assessment, approved in December 2012. Those objectives include ensuring Navy forces are capable and ready, contributing to the safety, stability and security in the region, safeguarding U.S. maritime interests, protecting critical infrastructure and key resources in the Arctic, and strengthening and fostering new cooperative relationships in the region. As a global Nation, the United States needs to ensure that the Navy is adequately prepared to preserve U.S. national security interests and collaborate with other Arctic nations if and when the region will be open for passage with increased traffic. The committee recognizes the importance of transparency of action without seeking to militarize the region. Therefore, the committee directs the Secretary of the Navy to provide a roadmap for future activities and costs for training and operating in the Arctic. This roadmap should be a derivative of the National Security Strategy, and shall identify proposed exercises (including table top exercises), to include the frequency, cost, and a more detailed investment strategy across the Program Objective Memorandum through fiscal years 2020 to support a timeframe leading to increased operations in the region between 2020 to 2030. Additionally, the Navy should include details regarding international forums in which they participate. The committee further directs the Secretary of the Navy to provide this roadmap to the congressional defense committees by February 28, 2014. (Pages 130-131)

August 2013 Senate Appropriations Committee Report Language

The Senate Appropriations Committee, in its report (S.Rept. 113-85 of August 1, 2013) on S. 1429, the FY2014 DOD Appropriations Act, stated the following in a section of the report discussing the Navy's research and development account:

> *Arctic Deep Water Basing.*—The Committee commends the Department's efforts to investigate potential security and resource issues related to the opening of Arctic sea-routes. The Committee encourages the Department to continue researching a range of solutions and response, including Arctic forward basing proposals. The Committee requests the Department to provide a report to the congressional defense committees on current and future programs related to Arctic security solutions and provide an analysis of procurement and costs related to both fixed and mobile Arctic basing options. (Page 163)

Coast Guard High Latitude Study Provided to Congress in July 2011

In July 2011, the Coast Guard provided to Congress a study on the Coast Guard's missions and capabilities for operations in high-latitude (i.e., polar) areas. The study, commonly known as the High Latitude Study, is dated July 2010 on its cover. The High Latitude Study concluded the following:

> [The study] concludes that future [Coast Guard] capability and capacity gaps will significantly impact four [Coast Guard] mission areas in the Arctic: Defense Readiness, Ice Operations, Marine Environmental Protection, and Ports, Waterways, and Coastal Security. These mission areas address the protection of important national interests in a geographic area where other nations are actively pursuing their own national goals. U.S. national policy and laws define the requirements to assert the nation's jurisdiction over its territory and interests; to ensure the security of its people and critical infrastructure; to participate fully in the collection of scientific knowledge; to support commercial enterprises with public utility; and to ensure that the Arctic environment is not degraded by increased human activity.
>
> The Coast Guard's ability to support Defense Readiness mission requirements in the Arctic is closely linked to DoD responsibilities. The Coast Guard presently possesses the only surface vessels capable of operating in ice-covered and ice-diminished waters. The Coast Guard supports (1) DoD missions such as the resupply of Thule Air Base in Greenland and logistics support (backup) for McMurdo Station in Antarctica and (2) Department of State (DoS) directed Freedom of Navigation Operations. These unique Coast Guard capabilities have been

noted by the Joint Chiefs of Staff, the Navy's Task Force Climate Change, and the recently issued Naval Operations Concept 2010.

The common and dominant contributor to these significant mission impacts is the gap in polar icebreaking capability ...[277]

Other capability gaps contributing to the impact on Coast Guard ability to carry out its missions in the Arctic include:

- Communications System Capability – Continuous coverage along Alaska's West Coast, the Bering Strait, and throughout the North Slope is required for exchanging voice and data communications with Coast Guard units and other government and commercial platforms offshore.
- Forward Operating Locations - No suitable facilities currently exist on the North Slope or near the Bering Strait with facilities sufficient to support extended aircraft servicing and maintenance. Aircraft must travel long distances and expend significant time transiting to and from adequate facilities. This gap reduces on-scene presence and capability to support sustained operations in the region.
- Environmental response in ice-covered waters - The technology and procedures for assessment and mitigation measures for oil spills in ice-covered waters are not fully developed or tested.

Capability gaps in the Arctic region have moderate impacts on [the Coast Guard's] Aids to Navigation (AtoN), Search and Rescue (SAR), and Other Law Enforcement (OLE) missions. Both AtoN and SAR involve the safety of mariners and will gain more importance not only as commerce and tourism cause an increase in maritime traffic, but as U.S. citizens in northern Alaska face more unpredictable conditions. Performance of OLE will be increasingly necessary to ensure the integrity of U.S. living marine resources from outside pressures....

In addition to the assessment of polar icebreaking needs, the Arctic mission analysis examined a set of theoretical mixes (force packages) of Coast Guard assets consisting of icebreakers, their embarked helicopters, and deployment alternatives using aviation forward operating locations in Arctic Alaska....

All [six] of the force mixes [considered in the study] add assets to the existing Coast Guard Alaska Patrol consisting of (1) a high-endurance cutter (not an icebreaker) deployed in the Bering Sea carrying a short range recovery helicopter, and (2) medium range recovery helicopters located at Kodiak in the Gulf of Alaska, and seasonally deployed to locations in Cold Bay and St. Paul Island....

These force packages and associated risk assessment provide a framework for acquisition planning as the Coast Guard implements a strategy for closing the capability gaps. By first recapitalizing the aging icebreakers, the Coast Guard provides a foundation for buildout of these force mixes. In addition to the cost of the icebreakers, the force packages require investment in forward operating locations and in medium range helicopters. The mission analysis reports developed rough order-of-magnitude cost estimates for forward operating locations at approximately $36M [million] each and for helicopters at $9M each....

The analysis shows that the current Coast Guard deployment posture is not capable of effective response in northern Alaska and that response may be improved through a mix of deployed cutters, aircraft, and supporting infrastructure including forward operating locations and communications/navigation systems.[278]

May 2013 Coast Guard Arctic Strategy

On May 21, 2013, the Coast Guard released a strategy document for the Arctic.[279] The executive summary of the document states in part:

The U.S. Coast Guard, as the maritime component of the U.S. Department of Homeland Security (DHS), has specific statutory responsibilities in U.S. Arctic waters. This strategy outlines the ends, ways, and means for achieving strategic objectives in the Arctic over the next 10 years. The Coast Guard is responsible for ensuring safe, secure, and environmentally responsible maritime activity in U.S. Arctic waters. Our efforts must be accomplished in close coordination with DHS components, and involve facilitating commerce, managing borders, and improving resilience to disasters.

The Coast Guard's current suite of cutters, boats, aircraft, and shore infrastructure must meet a number of near-term mission demands. The Coast Guard employs mobile command and control platforms such as large cutters and ocean-going ice-strengthened buoy tenders, as well as seasonal air and communications capabilities through leased or deployable assets and facilities. These mobile and seasonal assets and facilities have proven to be important enablers for front-line priorities in the region, including search and rescue operations, securing the maritime border, collecting critical intelligence, responding to potential disasters, and protecting the marine environment....

Although winter sea travel is still severely limited due to extensive ice coverage across the region, recent summer and early autumn sea ice extent record lows have made seasonal maritime navigation more feasible. Economic development, in the forms of resource extraction, adventure tourism, and trans-Arctic shipping drives much of the current maritime activity in the region.

[Oil and gas exploration] activities [in the region] bring risk, which can be mitigated through appropriate maritime governance. Additionally, tourism is increasing rapidly in the Arctic. Due to undeveloped shore-based infrastructure, much of the increased tourism is expected to involve transportation via passenger vessel, further increasing near- and offshore activities in Arctic waters.

This document outlines three strategic objectives in the Arctic for the U.S. Coast Guard over the next 10 years:

- Improving Awareness
- Modernizing Governance
- Broadening Partnerships

Improving Awareness: Coast Guard operations require precise and ongoing awareness of activities in the maritime domain. Maritime awareness in the Arctic is currently restricted due to limited surveillance, monitoring, and information system capabilities. Persistent awareness enables identification of threats, information-sharing with front-line partners, and improved risk management. Improving awareness requires close collaboration within DHS, as well as with the Departments of State, Defense, Interior, the National Science Foundation and other stakeholders to enhance integration, innovation, and fielding of emerging technologies. The Intelligence Community and non-federal partners are also vital stakeholders.

Modernizing Governance: The concept of governance involves institutions, structures of authority, and capabilities necessary to oversee maritime activities while safeguarding national interests. Limited awareness and oversight challenge maritime sovereignty, including the protection of natural resources and control of maritime borders. The Coast Guard will work within its authorities to foster collective efforts, both domestically and internationally, to improve Arctic governance. In so doing, the Coast Guard will review its own institutions and regimes of governance to prepare for future missions throughout the Arctic.

Broadening Partnerships: Success in the Arctic requires a collective effort across both the public and private sectors. Such a collective effort must be inclusive of domestic regulatory regimes; international collaborative forums such as the Arctic Council, International Maritime Organization (IMO), and Inuit Circumpolar Council; domestic and international partnerships; and local engagements in Arctic communities focusing on training and volunteer service. Success in the Arctic also depends upon close intergovernmental cooperation to support national interests, including working closely within DHS, as well as with the Department of State, Department of Interior and other Federal partners as the U.S. prepares to assume Chairmanship of the Arctic Council in 2015.

Beyond these three strategic objectives, there are a number of additional factors that will position the Coast Guard for long-term success. These factors include building national awareness of the Arctic and its opportunities, strengthening maritime regimes, improving public-private relationships through a national concept of operations, seeking necessary authorities, and identifying future requirements and resources to shape trends favorably. This strategy outlines a number of priorities, ranging from capabilities and requirements to advances in science and technology that will facilitate our Nation's success in the region. Specifically, the strategy advocates to leverage the entire DHS enterprise and component capabilities to secure our borders, prevent terrorism, adapt to changing environmental conditions, enable community resilience and inform future policy.

Operating in the Arctic is not a new venture for the Coast Guard. However, adapting to changing conditions will require foresight, focus, and clear priorities. This strategy will ensure we attain the aim of safe, secure, and environmentally responsible maritime activity in the Arctic by improving awareness, modernizing governance, and broadening partnerships to ensure long-term success.[280]

A June 2013 press report stated:

Arctic sea lanes may be staying open longer and traffic in the frigid waters may be increasing, but the Coast Guard's new 10-year plan for Arctic missions won't send more Coasties to the region or establish a year-round presence, the service's top officer said.

The service's seasonal, mobile approach has been "tested and proven ... over the last several years," Commandant Adm. Bob Papp said when he released the Coast Guard's Arctic Strategy on May 21. "That will be our approach in the coming decade."

The service will not increase deployment time to the Arctic, continuing to rely on standard six-month cutter deployments to maintain a presence, Papp said.

The national security cutters carry supplies, small boats and helicopters, and have the communication and intelligence capabilities needed for a sustained offshore presence in the ice-free waters, he said. They patrol the Arctic for the summer season, typically from June to September, depending on how long the waters are ice-free, said Lt. Veronica Colbath, a spokeswoman for District 17, which covers Alaska.

Using the cutter fleet instead of establishing bases over the next 10 years will save money and allow for greater flexibility, Papp said. The strategy also stresses expanding partnerships with domestic and international regulatory bodies monitoring the Arctic, including the Arctic Council, International Maritime Organization and the Inuit Circumpolar Council.

With long-term forecasts showing the region will become ice-free for longer periods, Papp said that at some point the Coast Guard might need a full-time sector, based in Barrow, Alaska.

However, that's not part of the current strategy.

"We are not going to permanently station anyone up there," Papp said.[281]

July 2013 Senate Appropriations Committee Report Language

The Senate Appropriations Committee, in its report (S.Rept. 113-77 of July 18, 2013) on H.R. 2217, the FY14 Department of Homeland Security Appropriations Act, stated:

> INFRASTRUCTURE AND RESPONSE CAPABILITIES IN THE ARCTIC
>
> The Committee is concerned about the lack of assets available for the Coast Guard's Arctic mission. No later than 120 days after the date of enactment of this act, the Commandant is directed to submit a report to the Committee comparing the costs of facility renovations to homeport and support an NSC [a National Security Cutter] in Alaska with the annual costs of transit time to Alaska area of operations for deployments and days lost to casualty repairs.
>
> ARCTIC STRATEGY
>
> The Coast Guard recently released its Arctic Strategy, which looks at the changing conditions in the region and contemplates future requirements. The strategy outlines three broad objectives: improving awareness; modernizing governance; and broadening partnerships. To carry out these objectives over the long term, the Coast Guard next needs to develop an implementation plan, including the identification of necessary capabilities, requirements, authorities, and resources. Therefore, the Coast Guard is directed to submit an Arctic strategy implementation plan no later than 120 days after the date of enactment of this act.
>
> INTERNATIONAL MARITIME ORGANIZATION
>
> The Committee wants to ensure that as activities in the Arctic expand, necessary response capabilities exists in the region. No later than 90 days after the date of enactment of this act, the Secretary is directed to submit a report to the Committee on activities that have occurred pursuant to section 307(b) of Public Law 111–281 and how the Department is meeting the requirements being developed in the agreements envisioned by that subsection.282 (Pages 80- 81)

U.S. Military Cooperation with Canada and Other Countries

In December 2009, it was reported that "U.S. and Canadian defense officials are studying emerging gaps in their awareness of Arctic activities, seeking to boost North American Aerospace Defense [NORAD] Command's maritime-warning mission and crafting a new threat assessment for the region." The effort would reportedly involve both NORAD and a Canadian-American advisory board called the Permanent Joint Board on Defense.[283]

In May 2010, it was reported that "American and Canadian defense officials are bolstering collaboration on military exercises, investment plans and technology development related to the Arctic." The report stated that "U.S. and Canadian intelligence officials have created a classified joint "utilization" assessment for the Arctic looking out to 2020, which will be continuously updated." It also stated that "American and Canadian defense officials have just started implementing a new five-year work plan that contains specific initiatives and requires concrete deliverables, Stockton said, noting the Arctic received special attention in the plan. The plan also covers defense critical infrastructure protection, defense support to civil agencies and defense cooperation in the Americas."[284]

In December 2012, the commander of NORAD and USNORTHCOM and the commander of the Canadian Joint Operations Command signed two documents on Arctic cooperation. A December 11, 2012, statement issued by NORAD and USNORTHCOM on the signing of the documents stated in part:

> Army Gen. Charles Jacoby, Jr., commander of NORAD and U.S. Northern Command, and Lt.-Gen. Stuart Beare, Canadian Joint Operations Command commander, signed two significant documents today—the Tri Command Framework for Arctic Cooperation and the Tri-Command Training and Exercise Statement of Intent—during the 230th meeting of the Canada-U.S. Permanent Joint Board on Defense in Colorado Springs, Colo.
>
> The Tri-Command Framework for Arctic Cooperation acknowledges the Arctic is not a region of conflict and the Canadian and U.S. militaries will support other departments and agencies in response to threats and hazards in the region when requested and directed. In that context, the goal of the Framework is to promote enhanced military cooperation in the Arctic and identify specific areas of potential Tri-Command cooperation in the preparation for and conduct of safety, security and defense operations. It strengthens an already unique and mature partnership where coordination and cooperation occurs on a regular basis. The Tri-Command Framework for Arctic Cooperation document is not a plan but rather outlines a process that supports the identification of opportunities for potential cooperation in the Arctic. Areas that continue to be improved, particularly in the Arctic, include planning, domain awareness, information-sharing, training and exercises, operations, capability development, and science and technology....
>
> The second document, the Tri Command Training and Exercise Statement of Intent, is aimed at enhancing joint and combined readiness

in support of safety, security and defence missions through combined training and exercises and reinforcing partnerships and collaboration among the Commands. The Tri-Command Training and Exercise Directive will help to ensure a timely and coordinated response to safety, security and defence challenges to North America.

The signing of the Tri Command Framework for Arctic Cooperation and the Tri-Command Training and Exercise Statement of Intent follows on a Tri Command Strategy that spells out a series of shared tasks designed to strengthen working relationships with all Tri-Command defense and security partners.[285]

In June 2013, it was reported that

Defense chiefs representing the world's eight main Arctic nations will strengthen cooperation in marine surveillance and expand joint military exercises.

Moreover, defense commanders agreed to identify and appraise the military and civilian capabilities in each country that can be used to support civilian missions in the Arctic over the next 12 months.

The new strategy, following a two-day meeting of defense commanders in the coastal Greenland town of Ilulissat that ended June 12, will focus on how the eight Arctic nations can bolster defense and security cooperation in the Arctic and how military resources can be better deployed to support civilian needs across borders.[286]

January 2009 Arctic Policy Directive (NSPD 66/HSPD 25)

As noted earlier (see "January 2009 Arctic Policy Directive" in "Background"), the Obama Administration is currently operating under the January 2009 Arctic region policy directive (NSPD 66/HSPD 25) issued by the George W. Bush Administration.[287] Potential oversight questions include but not are limited to the following:

- Is NSPD 66/HSPD 25 a suitable statement of U.S. policy for the Arctic region?[288]
- Although the Obama Administration is currently operating under NSPD 66/HSPD 25, does the Obama Administration fully agree with all parts of it? If not, with which parts does it not fully agree?
- Does the Obama Administration intend to eventually conduct a review of NSDP 66/HSPD 25? If so, what is the Administration's schedule for conducting and releasing the results of that review?

In connection with the above questions, a March 2013 report from the Center for Strategic and International Studies (CSIS) on U.S. interests and U.S. government actors in the Arctic stated:

> NSPD-66/HSPD-25 contains a rigorous implementation schedule for the federal government. Yet, as this directive passes its fourth anniversary, many implementation strategies have not been put into effect. That said, several of the policy priorities and implementation strategies are redundant. For example, to meet U.S. security interests in the Arctic, an interagency group under the leadership of the departments of State, Defense, and Homeland Security is requested to "develop greater capabilities and capacity [...] to protect U.S. air, land and sea borders" and "increase Arctic maritime domain awareness." Under maritime transportation, these same agencies, with the departments of Transportation and Commerce, are to "determine basing and logistics support requirements, including necessary airlift and icebreaking capabilities; and improve plans and cooperative agreements for search and rescue." These overlapping mandates raise the question whether this strategy is a security function, an act of commerce—or both—and who ultimately is accountable for implementing the strategy.
>
> Such policy overlap and redundancy accurately reflects U.S. Arctic policymaking today and unfortunately reflects poorly on the interagency process following the release of NSPD-66. A successful future policy demands a more streamlined and prioritized process.
>
> It is time to update NSPD-66. Other Arctic coastal states emerged with Arctic policy statements around the same time as the United States did, in the 2007 to 2009 timeframe. Since issuing these statements, however, most other Arctic nations have updated and further refined their whole-of-government strategies. Unfortunately, the United States has failed to do so. Separate federal agencies and departments have developed, or will be developing, their own separate strategies based on NSPD-66, but this effort has been uneven at best ... The Department of Defense report on Arctic Operations and the Northwest Passage was congressionally mandated by the fiscal year 2011 National Defense Authorization Act. However, there is no such mandate for other U.S. government agencies and, consequently, most other agencies have yet to develop their own implementation strategy stemming from NSPD-66.
>
> What would an updated U.S. Arctic strategy require? The seven policy areas identified in NSPD-66 remain relevant, but these areas must be defined more clearly.
>
> First and foremost, the United States must create a long-term economic strategy for the American Arctic. The first component of an Arctic economic strategy must be an energy, mineral resource, and infrastructure strategy. NSPD-66 states that "Energy development in the

Arctic region will play an important role in meeting growing global energy demand." How large a role? At present, U.S. energy strategy consists of a five-year offshore licensing and permitting plan that ends in 2017. The government needs to define its long-term offshore and onshore energy strategy for Alaska, answering such questions as whether America's Arctic energy resources are intended to meet U.S. demand or to be exported to Asian markets. Other questions in need of answers include the status of port, pipeline, and liquid natural gas infrastructure; whether methane hydrates are viable; and whether energy development can be pursued in an environmentally sustainable way in such a fragile environment.

Based on the answers to these questions, America's Arctic economic strategy must also include a detailed maritime transportation and infrastructure strategy. This strategy would ideally be built around ecosystem-based management. It is clear that a future U.S. Arctic maritime or infrastructure strategy will likely be a public-private partnership, as the private sector will provide significantly more financial and physical resources than will the U.S. government.[289]

APPENDIX A. ARCTIC RESEARCH AND POLICY ACT (ARPA) OF 1984 (TITLE I OF P.L. 98-373)

The text of the Arctic Research and Policy Act (ARPA) of 1984 (Title I of P.L. 98-373 of July 31, 1984)[290] is as follows:

TITLE I – ARCTIC RESEARCH AND POLICY

SHORT TITLE

SEC. 101. This title may be cited as the "Arctic Research and Policy Act of 1984".

FINDINGS AND PURPOSES

SEC. 102.

(a) The Congress finds and declares that-

(1) the Arctic, onshore and offshore, contains vital energy resources that can reduce the Nation's dependence on foreign oil and improve the national balance of payments;

(2) as the Nation's only common border with the Soviet Union, the Arctic is critical to national defense;

(3) the renewable resources of the Arctic, specifically fish and other seafood, represent one of the Nation's greatest commercial assets;

(4) Arctic conditions directly affect global weather patterns and must be understood in order to promote better agricultural management throughout the United States;
(5) industrial pollution not originating in the Arctic region collects in the polar air mass, has the potential to disrupt global weather patterns, and must be controlled through international cooperation and consultation;
(6) the Arctic is a natural laboratory for research into human health and adaptation, physical and psychological, to climates of extreme cold and isolation and may provide information crucial for future defense needs;
(7) atmospheric conditions peculiar to the Arctic make the Arctic a unique testing ground for research into high latitude communications, which is likely to be crucial for future defense needs;
(8) Arctic marine technology is critical to cost-effective recovery and transportation of energy resources and tc the national defense;
(9) the United States has important security, economic, and environmental interests in developing and maintaining a fleet of icebreaking vessels capable of operating effectively in the heavy ice regions of the Arctic;
(10)most Arctic-rim countries, particularly the Soviet Union, possess Arctic technologies far more advanced than those currently available in the United States;
(11)Federal Arctic research is fragmented and uncoordinated at the present time, leading to the neglect of certain areas of research and to unnecessary duplication of effort in other areas of research;
(12)improved logistical coordination and support for Arctic research and better dissemination of research data and information is necessary to increase the efficiency and utility of national Arctic research efforts;
(13)a comprehensive national policy and program plan to organize and fund currently neglected scientific research with respect to the Arctic is necessary to fulfill national objectives in Arctic research;
(14)the Federal Government, in cooperation with State and local governments, should focus its efforts on the collection and characterization of basic data related to biological, materials, geophysical, social, and behavioral phenomena in the Arctic;

(15) research into the long-range health, environmental, and social effects of development in the Arctic is necessary to mitigate the adverse consequences of that development to the land and its residents;

(16) Arctic research expands knowledge of the Arctic, which can enhance the lives of Arctic residents, increase opportunities for international cooperation among Arctic-rim countries, and facilitate the formulation of national policy for the Arctic; and

(17) the Alaskan Arctic provides an essential habitat for marine mammals, migratory waterfowl, and other forms of wildlife which are important to the Nation and which are essential to Arctic residents.

(b) The purposes of this title are-

(1) to establish national policy, priorities, and goals and to provide a Federal program plan for basic and applied scientific research with respect to the Arctic, including natural resources and materials, physical, biological and health sciences, and social and behavioral sciences;

(2) to establish an Arctic Research Commission to promote Arctic research and to recommend Arctic research policy;

(3) to designate the National Science Foundation as the lead agency responsible for implementing Arctic research policy; and

(4) to establish an Interagency Arctic Research Policy Committee to develop a national Arctic research policy and a five year plan to implement that policy.

ARCTIC RESEARCH COMMISSION
SEC. 103.

(a) The President shall establish an Arctic Research Commission (hereafter referred to as the "Commission").

(b)

(1) The Commission shall be composed of five members appointed by the President, with the Director of the National Science Foundation serving as a nonvoting, ex officio member. The members appointed by the President shall include-

(A) three members appointed from among individuals from academic or other research institutions with expertise in areas of research relating to the Arctic, including the physical, biological, health, environmental, social, and behavioral sciences;

(B) one member appointed from among indigenous residents of the Arctic who are representative of the needs and interests of Arctic residents and who live in areas directly affected by Arctic resource development; and

(C) one member appointed from among individuals familiar with the Arctic and representative of the needs and interests of private industry undertaking resource development in the Arctic.

(2) The President shall designate one of the appointed members of the Commission to be chairperson of the Commission.

(c)

(1) Except as provided in paragraph (2) of this subsection, the term of office of each member of the Commission appointed under subsection (b)(1) shall be four years.

(2) Of the members of the Commission originally appointed under subsection (b)(1)-

(A) one shall be appointed for a term of two years;

(B) two shall be appointed for a term of three years; and

(C) two shall be appointed for a term of four years.

(3) Any vacancy occurring in the membership of the Commission shall be filled, after notice of the vacancy is published in the Federal Register, in the manner provided by the preceding provisions of this section, for the remainder of the unexpired term.

(4) A member may serve after the expiration of the member's term of office until the President appoints a successor.

(5) A member may serve consecutive terms beyond the member's original appointment.

(d)

(1) Members of the Commission may be allowed travel expenses, including per diem in lieu of subsistence, as authorized by section 5703 of title 5, United States Code. A member of the Commission not presently employed for compensation shall be compensated at a rate equal to the daily equivalent of the rate for GS-16 of the General Schedule under section 5332 of title 5, United States Code, for each day the member is engaged in the actual performance of his duties as a member of the Commission, not to

exceed 90 days of service each year. Except for the purposes of chapter 81 of title 5 (relating to compensation for work injuries) and chapter 171 of title 28 (relating to tort claims), a member of the Commission shall not be considered an employee of the United States for any purpose.

(2) The Commission shall meet at the call of its Chairman or a majority of its members.

(3) Each Federal agency referred to in section 107(b) may designate a representative to participate as an observer with the Commission. These representatives shall report to and advise the Commission on the activities relating to Arctic research of their agencies.

(4) The Commission shall conduct at least one public meeting in the State of Alaska annually.

DUTIES OF COMMISSION
SEC. 104.

(a) The Commission shall-

(1) develop and recommend an integrated national Arctic research policy;

(2) in cooperation with the Interagency Arctic Research Policy Committee established under section 107, assist in establishing a national Arctic research program plan to implement the Arctic research policy;

(3) facilitate cooperation between the Federal Government and State and local governments with respect to Arctic research;

(4) review Federal research programs in the Arctic and suggest improvements in coordination among programs;

(5) recommend methods to improve logistical planning and support for Arctic research as may be appropriate and in accordance with the findings and purposes of this title;

(6) suggest methods for improving efficient sharing and dissemination of data and information on the Arctic among interested public and private institutions;

(7) offer other recommendations and advice to the Interagency Committee established under section 107 as it may find appropriate; and

(8) cooperate with the Governor of the State of Alaska and with agencies and organizations of that State which the Governor may

designate with respect to the formulation of Arctic research policy.

(b) Not later than January 31 of each year, the Commission shall-
  (1) publish a statement of goals and objectives with respect to Arctic research to guide the Interagency Committee established under section 107 in the performance of its duties; and
  (2) submit to the President and to the Congress a report describing the activities and accomplishments of the Commission during the immediately preceding fiscal year.

COOPERATION WITH THE COMMISSION
SEC. 105.

(a)
  (1) The Commission may acquire from the head of any Federal agency unclassified data, reports, and other nonproprietary information with respect to Arctic research in the possession of the agency which the Commission considers useful in the discharge of its duties.
  (2) Each agency shall cooperate with the Commission and furnish all data, reports, and other information requested by the Commission to the extent permitted by law; except that no agency need furnish any information which it is permitted to withhold under section 552 of title 5, United States Code.

(b) With the consent of the appropriate agency head, the Commission may utilize the facilities and services of any Federal agency to the extent that the facilities and services are needed for the establishment and development of an Arctic research policy, upon reimbursement to be agreed upon by the Commission and the agency head and taking every feasible step to avoid duplication of effort.

(c) All Federal agencies shall consult with the Commission before undertaking major Federal actions relating to Arctic research.

ADMINISTRATION OF THE COMMISSION
SEC. 106. The Commission may-

(1) in accordance with the civil service laws and subchapter III of chapter 53 of title 5, United States Code, appoint and fix the compensation of an Executive Director and necessary additional staff personnel, but not to exceed a total of seven compensated personnel;

(2) procure temporary and intermittent services as authorized by section 3109 of title 5, United States Code;
(3) enter into contracts and procure supplies, services, and personal property; and
(4) enter into agreements with the General Services Administration for the procurement of necessary financial and administrative services, for which payment shall be made by reimbursement from funds of the Commission in amounts to be agreed upon by the Commission and the Administrator of the General Services Administration.

LEAD AGENCY AND INTERAGENCY ARCTIC RESEARCH POLICY COMMITTEE

SEC. 107.

(a) The National Science Foundation is designated as the lead agency responsible for implementing Arctic research policy, and the Director of the National Science Foundation shall insure that the requirements of section 108 are fulfilled.

(b)
   (1) The President shall establish an Interagency Arctic Research Policy Committee (hereinafter referred to as the "Interagency Committee").
   (2) The Interagency Committee shall be composed of representatives of the following Federal agencies or offices:
      (A) the National Science Foundation;
      (B) the Department of Commerce;
      (C) the Department of Defense;
      (D) the Department of Energy;
      (E) the Department of the Interior;
      (F) the Department of State;
      (G) the Department of Transportation;
      (H) the Department of Health and Human Services;
      (I) the National Aeronautics and Space Administration;
      (J) the Environmental Protection Agency; and
      (K) any other agency or office deemed appropriate.
   (3) The representative of the National Science Foundation shall serve as the Chairperson of the Interagency Committee.

DUTIES OF THE INTERAGENCY COMMITTEE

SEC. 108.

(a) The Interagency Committee shall-

(1) survey Arctic research conducted by Federal, State, and local agencies, universities, and other public and private institutions to help determine priorities for future Arctic research, including natural resources and materials, physical and biological sciences, and social and behavioral sciences;

(2) work with the Commission to develop and establish an integrated national Arctic research policy that will guide Federal agencies in developing and implementing their research programs in the Arctic;

(3) consult with the Commission on-

(A) the development of the national Arctic research policy and the 5-year plan implementing the policy;

(B) Arctic research programs of Federal agencies;

(C) recommendations of the Commission on future Arctic research; and

(D) guidelines for Federal agencies for awarding and administering Arctic research grants;

(4) develop a 5-year plan to implement the national policy, as provided for in section 109;

(5) provide the necessary coordination, data, and assistance for the preparation of a single integrated, coherent, and multiagency budget request for Arctic research as provided for in section 110;

(6) facilitate cooperation between the Federal Government and State and local governments in Arctic research, and recommend the undertaking of neglected areas of research in accordance with the findings and purposes of this title;

(7) coordinate and promote cooperative Arctic scientific research programs with other nations, subject to the foreign policy guidance of the Secretary of State;

(8) cooperate with the Governor of the State of Alaska in fulfilling its responsibilities under this title;

(9) promote Federal interagency coordination of all Arctic research activities, including-

(A) logistical planning and coordination; and

(B) the sharing of data and information associated with Arctic research, subject to section 552 of title 5, United States Code; and

(10) provide public notice of its meetings and an opportunity for the public to participate in the development and implementation of national Arctic research policy.

(b) Not later than January 31, 1986, and biennially thereafter, the Interagency Committee shall submit to the Congress through the President, a brief, concise report containing-
   (1) a statement of the activities and accomplishments of the Interagency Committee since its last report; and
   (2) a description of the activities of the Commission, detailing with particularity the recommendations of the Commission with respect to Federal activities in Arctic research.

5-YEAR ARCTIC RESEARCH PLAN
SEC. 109.

(a) The Interagency Committee, in consultation with the Commission, the Governor of the State of Alaska, the residents of the Arctic, the private sector, and public interest groups, shall prepare a comprehensive 5-year program plan (hereinafter referred to as the "Plan") for the overall Federal effort in Arctic research. The Plan shall be prepared and submitted to the President for transmittal to the Congress within one year after the enactment of this Act and shall be revised biennially thereafter.

(b) The Plan shall contain but need not be limited to the following elements:
   (1) an assessment of national needs and problems regarding the Arctic and the research necessary to address those needs or problems;
   (2) a statement of the goals and objectives of the Interagency Committee for national Arctic research;
   (3) a detailed listing of all existing Federal programs relating to Arctic research, including the existing goals, funding levels for each of the 5 following fiscal years, and the funds currently being expended to conduct the programs;
   (4) recommendations for necessary program changes and other proposals to meet the requirements of the policy and goals as set forth by the Commission and in the Plan as currently in effect; and

(5) a description of the actions taken by the Interagency Committee to coordinate the budget review process in order to ensure interagency coordination and cooperation in (A) carrying out Federal Arctic research programs, and (B) eliminating unnecessary duplication of effort among these programs.

COORDINATION AND REVIEW OF BUDGET REQUESTS
SEC. 110.

(a) The Office of Science and Technology Policy shall-
   (1) review all agency and department budget requests related to the Arctic transmitted pursuant to section 108(a)(5), in accordance with the national Arctic research policy and the 5-year program under section 108(a)(2) and section 109, respectively; and
   (2) consult closely with the Interagency Committee and the Commission to guide the Office of Science and Technology Policy's efforts.

(b)
   (1) The Office of Management and Budget shall consider all Federal agency requests for research related to the Arctic as one integrated, coherent, and multiagency request which shall be reviewed by the Office of Management and Budget prior to submission of the President's annual budget request for its adherence to the Plan. The Commission shall, after submission of the President's annual budget request, review the request and report to Congress on adherence to the Plan.
   (2) The Office of Management and Budget shall seek to facilitate planning for the design, procurement, maintenance, deployment, and operations of icebreakers needed to provide a platform for Arctic research by allocating all funds necessary to support icebreaking operations, except for recurring incremental costs associated with specific projects, to the Coast Guard.

AUTHORIZATION OF APPROPRIATIONS; NEW SPENDING AUTHORITY
SEC. 111.

(a) There are authorized to be appropriated such sums as may be necessary for carrying out this title.

(b) Any new spending authority (within the meaning of section 401 of the Congressional Budget Act of 1974) which is provided under this title shall be effective for any fiscal year only to such extent or in such amounts as may be provided in appropriation Acts.

DEFINITION

SEC. 112. As used in this title, the term "Arctic" means all United States and foreign territory north of the Arctic Circle and all United States territory north and west of the boundary formed by the Porcupine, Yukon, and Kuskokwim Rivers; all contiguous seas, including the Arctic Ocean and the Beaufort, Bering, and Chukchi Seas; and the Aleutian chain.

APPENDIX B. P.L. 101-609 OF 1990, AMENDING ARCTIC RESEARCH AND POLICY ACT (ARPA) OF 1984

The Arctic Research and Policy Act (ARPA) of 1984 (see Appendix A) was amended by P.L. 101-609 of November 16, 1990. The text of P.L. 101-609 is as follows:

SECTION 1. Except as specifically provided in this Act, whenever in this Act an amendment or repeal is expressed as an amendment to, or repeal of a provision, the reference shall be deemed to be made to the Arctic Research and Policy Act of 1984.

SEC. 2. Section 103(b)(1) (15 U.S.C. 4102(b)(1)) is amended—

(1) in the text above clause (A), by striking out `five' and inserting in lieu thereof `seven';
(2) in clause (A), by striking out `three' and inserting in lieu thereof `four'; and
(3) in clause (C), by striking out `one member' and inserting in lieu thereof `two members'.

SEC. 3. Section 103(d)(1) (15 U.S.C. 4102(d)(1)) is amended by striking out `GS-16' and inserting in lieu thereof `GS-18'.

SEC. 4.

(a) Section 104(a) (15 U.S.C. 4102(a)) is amended—

(1) in paragraph (4), by striking out `suggest' and inserting in lieu thereof `recommend';
(2) in paragraph (6), by striking out `suggest' and inserting in lieu thereof `recommend';
(3) in paragraph (7), by striking out `and' at the end thereof;
(4) in paragraph (8), by striking out the period and inserting in lieu thereof a semicolon; and
(5) by adding at the end thereof the following new paragraphs:
 '(9) recommend to the Interagency Committee the means for developing international scientific cooperation in the Arctic; and
 '(10) not later than January 31, 1991, and every 2 years thereafter, publish a statement of goals and objectives with respect to Arctic research to guide the Interagency Committee established under section 107 in the performance of its duties.'.

(b) Section 104(b) is amended to read as follows:
 '(b) Not later than January 31 of each year, the Commission shall submit to the President and to the Congress a report describing the activities and accomplishments of the Commission during the immediately preceding fiscal year.'.

SEC. 5. Section 106 (15 U.S.C. 4105) is amended—

(1) in paragraph (3), by striking out `and' at the end thereof;
(2) in paragraph (4), by striking out the period at the end thereof and inserting in lieu thereof; and'; and
(3) by adding at the end thereof the following new paragraph:
 '(5) appoint, and accept without compensation the services of, scientists and engineering specialists to be advisors to the Commission. Each advisor may be allowed travel expenses, including per diem in lieu of subsistence, as authorized by section 5703 of title 5, United States Code. Except for the purposes of chapter 81 of title 5 (relating to compensation for work injuries) and chapter 171 of title 28 (relating to tort claims) of the United States Code, an advisor appointed under this paragraph shall not be considered an employee of the United States for any purpose.'

SEC. 6. Subsection (b)(2) of section 108 (15 U.S.C. 4107(b)(2)) is amended to read as follows:

'(2) a statement detailing with particularity the recommendations of the Commission with respect to Federal interagency activities in Arctic research and the disposition and responses to those recommendations.'

Appendix C. January 2009 Arctic Policy Directive (NSPD 66/HSPD 25)

On January 12, 2009, the George W. Bush Administration released a presidential directive establishing a new U.S. policy for the Arctic region. The directive, dated January 9, 2009, was issued as National Security Presidential Directive 66/Homeland Security Presidential Directive 25 (NSPD 66/HSPD 25). The text of NSPD 66/HSPD 25 is as follows:

SUBJECT: Arctic Region Policy

I. PURPOSE
   A. This directive establishes the policy of the United States with respect to the Arctic region and directs related implementation actions. This directive supersedes Presidential Decision Directive/NSC-26 (PDD-26; issued 1994) with respect to Arctic policy but not Antarctic policy; PDD-26 remains in effect for Antarctic policy only.
   B. This directive shall be implemented in a manner consistent with the Constitution and laws of the United States, with the obligations of the United States under the treaties and other international agreements to which the United States is a party, and with customary international law as recognized by the United States, including with respect to the law of the sea.

II. BACKGROUND
   A. The United States is an Arctic nation, with varied and compelling interests in that region. This directive takes into account several developments, including, among others:
      1. Altered national policies on homeland security and defense;
      2. The effects of climate change and increasing human activity in the Arctic region;
      3. The establishment and ongoing work of the Arctic Council; and

4. A growing awareness that the Arctic region is both fragile and rich in resources.

III. POLICY

A. It is the policy of the United States to:

1. Meet national security and homeland security needs relevant to the Arctic region;
2. Protect the Arctic environment and conserve its biological resources;
3. Ensure that natural resource management and economic development in the region are environmentally sustainable;
4. Strengthen institutions for cooperation among the eight Arctic nations (the United States, Canada, Denmark, Finland, Iceland, Norway, the Russian Federation, and Sweden);
5. Involve the Arctic's indigenous communities in decisions that affect them; and
6. Enhance scientific monitoring and research into local, regional, and global environmental issues.

B. National Security and Homeland Security Interests in the Arctic

1. The United States has broad and fundamental national security interests in the Arctic region and is prepared to operate either independently or in conjunction with other states to safeguard these interests. These interests include such matters as missile defense and early warning; deployment of sea and air systems for strategic sealift, strategic deterrence, maritime presence, and maritime security operations; and ensuring freedom of navigation and overflight.
2. The United States also has fundamental homeland security interests in preventing terrorist attacks and mitigating those criminal or hostile acts that could increase the United States vulnerability to terrorism in the Arctic region.
3. The Arctic region is primarily a maritime domain; as such, existing policies and authorities relating to maritime areas continue to apply, including those relating to law enforcement.[1] Human activity in the Arctic region is increasing and is projected to increase further in coming years. This requires the United States to assert a more active and influential national presence to protect its Arctic interests and to project sea power throughout the region.

4. The United States exercises authority in accordance with lawful claims of United States sovereignty, sovereign rights, and jurisdiction in the Arctic region, including sovereignty within the territorial sea, sovereign rights and jurisdiction within the United States exclusive economic zone and on the continental shelf, and appropriate control in the United States contiguous zone.
5. Freedom of the seas is a top national priority. The Northwest Passage is a strait used for international navigation, and the Northern Sea Route includes straits used for international navigation; the regime of transit passage applies to passage through those straits. Preserving the rights and duties relating to navigation and overflight in the Arctic region supports our ability to exercise these rights throughout the world, including through strategic straits.
6. Implementation: In carrying out this policy as it relates to national security and homeland security interests in the Arctic, the Secretaries of State, Defense, and Homeland Security, in coordination with heads of other relevant executive departments and agencies, shall:
    a. Develop greater capabilities and capacity, as necessary, to protect United States air, land, and sea borders in the Arctic region;
    b. Increase Arctic maritime domain awareness in order to protect maritime commerce, critical infrastructure, and key resources;
    c. Preserve the global mobility of United States military and civilian vessels and aircraft throughout the Arctic region;
    d. Project a sovereign United States maritime presence in the Arctic in support of essential United States interests; and
    e. Encourage the peaceful resolution of disputes in the Arctic region.

C. International Governance

1. The United States participates in a variety of fora, international organizations, and bilateral contacts that promote United States interests in the Arctic. These include the Arctic Council, the International Maritime Organization

(IMO), wildlife conservation and management agreements, and many other mechanisms. As the Arctic changes and human activity in the region increases, the United States and other governments should consider, as appropriate, new international arrangements or enhancements to existing arrangements.

2. The Arctic Council has produced positive results for the United States by working within its limited mandate of environmental protection and sustainable development. Its subsidiary bodies, with help from many United States agencies, have developed and undertaken projects on a wide range of topics. The Council also provides a beneficial venue for interaction with indigenous groups. It is the position of the United States that the Arctic Council should remain a high-level forum devoted to issues within its current mandate and not be transformed into a formal international organization, particularly one with assessed contributions. The United States is nevertheless open to updating the structure of the Council, including consolidation of, or making operational changes to, its subsidiary bodies, to the extent such changes can clearly improve the Council's work and are consistent with the general mandate of the Council.
3. The geopolitical circumstances of the Arctic region differ sufficiently from those of the Antarctic region such that an "Arctic Treaty" of broad scope—along the lines of the Antarctic Treaty—is not appropriate or necessary.
4. The Senate should act favorably on U.S. accession to the U.N. Convention on the Law of the Sea promptly, to protect and advance U.S. interests, including with respect to the Arctic. Joining will serve the national security interests of the United States, including the maritime mobility of our Armed Forces worldwide. It will secure U.S. sovereign rights over extensive marine areas, including the valuable natural resources they contain. Accession will promote U.S. interests in the environmental health of the oceans. And it will give the United States a seat at the table when the rights that are vital to our interests are debated and interpreted.
5. Implementation: In carrying out this policy as it relates to international governance, the Secretary of State, in

coordination with heads of other relevant executive departments and agencies, shall:

a. Continue to cooperate with other countries on Arctic issues through the United Nations (U.N.) and its specialized agencies, as well as through treaties such as the U.N. Framework Convention on Climate Change, the Convention on International Trade in Endangered Species of Wild Fauna and Flora, the Convention on Long Range Transboundary Air Pollution and its protocols, and the Montreal Protocol on Substances that Deplete the Ozone Layer;
b. Consider, as appropriate, new or enhanced international arrangements for the Arctic to address issues likely to arise from expected increases in human activity in that region, including shipping, local development and subsistence, exploitation of living marine resources, development of energy and other resources, and tourism;
c. Review Arctic Council policy recommendations developed within the ambit of the Council's scientific reviews and ensure the policy recommendations are subject to review by Arctic governments; and
d. Continue to seek advice and consent of the United States Senate to accede to the 1982 Law of the Sea Convention.

D. Extended Continental Shelf and Boundary Issues

1. Defining with certainty the area of the Arctic seabed and subsoil in which the United States may exercise its sovereign rights over natural resources such as oil, natural gas, methane hydrates, minerals, and living marine species is critical to our national interests in energy security, resource management, and environmental protection. The most effective way to achieve international recognition and legal certainty for our extended continental shelf is through the procedure available to States Parties to the U.N. Convention on the Law of the Sea.
2. The United States and Canada have an unresolved boundary in the Beaufort Sea. United States policy recognizes a boundary in this area based on equidistance. The United States recognizes that the boundary area may contain oil, natural gas, and other resources.

3. The United States and Russia are abiding by the terms of a maritime boundary treaty concluded in 1990, pending its entry into force. The United States is prepared to enter the agreement into force once ratified by the Russian Federation.
4. Implementation: In carrying out this policy as it relates to extended continental shelf and boundary issues, the Secretary of State, in coordination with heads of other relevant executive departments and agencies, shall:
   a. Take all actions necessary to establish the outer limit of the continental shelf appertaining to the United States, in the Arctic and in other regions, to the fullest extent permitted under international law;
   b. Consider the conservation and management of natural resources during the process of delimiting the extended continental shelf; and
   c. Continue to urge the Russian Federation to ratify the 1990 United States-Russia maritime boundary agreement.

E. Promoting International Scientific Cooperation
1. Scientific research is vital for the promotion of United States interests in the Arctic region. Successful conduct of U.S. research in the Arctic region requires access throughout the Arctic Ocean and to terrestrial sites, as well as viable international mechanisms for sharing access to research platforms and timely exchange of samples, data, and analyses. Better coordination with the Russian Federation, facilitating access to its domain, is particularly important.
2. The United States promotes the sharing of Arctic research platforms with other countries in support of collaborative research that advances fundamental understanding of the Arctic region in general and potential Arctic change in particular. This could include collaboration with bodies such as the Nordic Council and the European Polar Consortium, as well as with individual nations.
3. Accurate prediction of future environmental and climate change on a regional basis, and the delivery of near real-time information to end-users, requires obtaining, analyzing, and disseminating accurate data from the entire Arctic region, including both paleoclimatic data and observational data. The

United States has made significant investments in the infrastructure needed to collect environmental data in the Arctic region, including the establishment of portions of an Arctic circumpolar observing network through a partnership among United States agencies, academic collaborators, and Arctic residents. The United States promotes active involvement of all Arctic nations in these efforts in order to advance scientific understanding that could provide the basis for assessing future impacts and proposed response strategies.

4. United States platforms capable of supporting forefront research in the Arctic Ocean, including portions expected to be ice-covered for the foreseeable future, as well as seasonally ice-free regions, should work with those of other nations through the establishment of an Arctic circumpolar observing network. All Arctic nations are members of the Group on Earth Observations partnership, which provides a framework for organizing an international approach to environmental observations in the region. In addition, the United States recognizes that academic and research institutions are vital partners in promoting and conducting Arctic research.
5. Implementation: In carrying out this policy as it relates to promoting scientific international cooperation, the Secretaries of State, the Interior, and Commerce and the Director of the National Science Foundation, in coordination with heads of other relevant executive departments and agencies, shall:
   a. Continue to play a leadership role in research throughout the Arctic region;
   b. Actively promote full and appropriate access by scientists to Arctic research sites through bilateral and multilateral measures and by other means;
   c. Lead the effort to establish an effective Arctic circumpolar observing network with broad partnership from other relevant nations;
   d. Promote regular meetings of Arctic science ministers or research council heads to share information concerning scientific research opportunities and to improve coordination of international Arctic research programs;

e. Work with the Interagency Arctic Research Policy Committee (IARPC) to promote research that is strategically linked to U.S. policies articulated in this directive, with input from the Arctic Research Commission; and

f. Strengthen partnerships with academic and research institutions and build upon the relationships these institutions have with their counterparts in other nations.

F. Maritime Transportation in the Arctic Region

1. The United States priorities for maritime transportation in the Arctic region are:

   a. To facilitate safe, secure, and reliable navigation;

   b. To protect maritime commerce; and

   c. To protect the environment.

2. Safe, secure, and environmentally sound maritime commerce in the Arctic region depends on infrastructure to support shipping activity, search and rescue capabilities, short- and long- range aids to navigation, high-risk area vessel-traffic management, iceberg warnings and other sea ice information, effective shipping standards, and measures to protect the marine environment. In addition, effective search and rescue in the Arctic will require local, State, Federal, tribal, commercial, volunteer, scientific, and multinational cooperation.

3. Working through the International Maritime Organization (IMO), the United States promotes strengthening existing measures and, as necessary, developing new measures to improve the safety and security of maritime transportation, as well as to protect the marine environment in the Arctic region. These measures may include ship routing and reporting systems, such as traffic separation and vessel traffic management schemes in Arctic chokepoints; updating and strengthening of the Guidelines for Ships Operating in Arctic Ice- Covered Waters; underwater noise standards for commercial shipping; a review of shipping insurance issues; oil and other hazardous material pollution response agreements; and environmental standards.

4. Implementation: In carrying out this policy as it relates to maritime transportation in the Arctic region, the Secretaries

of State, Defense, Transportation, Commerce, and Homeland Security, in coordination with heads of other relevant executive departments and agencies, shall:

a. Develop additional measures, in cooperation with other nations, to address issues that are likely to arise from expected increases in shipping into, out of, and through the Arctic region;
b. Commensurate with the level of human activity in the region, establish a risk-based capability to address hazards in the Arctic environment. Such efforts shall advance work on pollution prevention and response standards; determine basing and logistics support requirements, including necessary airlift and icebreaking capabilities; and improve plans and cooperative agreements for search and rescue;
c. Develop Arctic waterways management regimes in accordance with accepted international standards, including vessel traffic-monitoring and routing; safe navigation standards; accurate and standardized charts; and accurate and timely environmental and navigational information; and
d. Evaluate the feasibility of using access through the Arctic for strategic sealift and humanitarian aid and disaster relief.

G. Economic Issues, Including Energy

1. Sustainable development in the Arctic region poses particular challenges. Stakeholder input will inform key decisions as the United States seeks to promote economic and energy security. Climate change and other factors are significantly affecting the lives of Arctic inhabitants, particularly indigenous communities. The United States affirms the importance to Arctic communities of adapting to climate change, given their particular vulnerabilities.
2. Energy development in the Arctic region will play an important role in meeting growing global energy demand as the area is thought to contain a substantial portion of the world's undiscovered energy resources. The United States seeks to ensure that energy development throughout the Arctic occurs in an environmentally sound manner, taking

into account the interests of indigenous and local communities, as well as open and transparent market principles. The United States seeks to balance access to, and development of, energy and other natural resources with the protection of the Arctic environment by ensuring that continental shelf resources are managed in a responsible manner and by continuing to work closely with other Arctic nations.

3. The United States recognizes the value and effectiveness of existing fora, such as the Arctic Council, the International Regulators Forum, and the International Standards Organization.
4. Implementation: In carrying out this policy as it relates to economic issues, including energy, the Secretaries of State, the Interior, Commerce, and Energy, in coordination with heads of other relevant executive departments and agencies, shall:
    a. Seek to increase efforts, including those in the Arctic Council, to study changing climate conditions, with a view to preserving and enhancing economic opportunity in the Arctic region. Such efforts shall include inventories and assessments of villages, indigenous communities, subsistence opportunities, public facilities, infrastructure, oil and gas development projects, alternative energy development opportunities, forestry, cultural and other sites, living marine resources, and other elements of the Arctic's socioeconomic composition;
    b. Work with other Arctic nations to ensure that hydrocarbon and other development in the Arctic region is carried out in accordance with accepted best practices and internationally recognized standards and the 2006 Group of Eight (G-8) Global Energy Security Principles;
    c. Consult with other Arctic nations to discuss issues related to exploration, production, environmental and socioeconomic impacts, including drilling conduct, facility sharing, the sharing of environmental data, impact assessments, compatible monitoring programs,

and reservoir management in areas with potentially shared resources;

d. Protect United States interests with respect to hydrocarbon reservoirs that may overlap boundaries to mitigate adverse environmental and economic consequences related to their development;
e. Identify opportunities for international cooperation on methane hydrate issues, North Slope hydrology, and other matters;
f. Explore whether there is a need for additional fora for informing decisions on hydrocarbon leasing, exploration, development, production, and transportation, as well as shared support activities, including infrastructure projects; and
g. Continue to emphasize cooperative mechanisms with nations operating in the region to address shared concerns, recognizing that most known Arctic oil and gas resources are located outside of United States jurisdiction.

H. Environmental Protection and Conservation of Natural Resources
   1. The Arctic environment is unique and changing. Increased human activity is expected to bring additional stressors to the Arctic environment, with potentially serious consequences for Arctic communities and ecosystems.
   2. Despite a growing body of research, the Arctic environment remains poorly understood. Sea ice and glaciers are in retreat. Permafrost is thawing and coasts are eroding. Pollutants from within and outside the Arctic are contaminating the region. Basic data are lacking in many fields. High levels of uncertainty remain concerning the effects of climate change and increased human activity in the Arctic. Given the need for decisions to be based on sound scientific and socioeconomic information, Arctic environmental research, monitoring, and vulnerability assessments are top priorities. For example, an understanding of the probable consequences of global climate variability and change on Arctic ecosystems is essential to guide the effective long-term management of Arctic natural resources and to address socioeconomic impacts of changing patterns in the use of natural resources.

3. Taking into account the limitations in existing data, United States efforts to protect the Arctic environment and to conserve its natural resources must be risk-based and proceed on the basis of the best available information.
4. The United States supports the application in the Arctic region of the general principles of international fisheries management outlined in the 1995 Agreement for the Implementation of the Provisions of the United Nations Convention on the Law of the Sea of December 10, 1982, relating to the Conservation and Management of Straddling Fish Stocks and Highly Migratory Fish Stocks and similar instruments. The United States endorses the protection of vulnerable marine ecosystems in the Arctic from destructive fishing practices and seeks to ensure an adequate enforcement presence to safeguard Arctic living marine resources.
5. With temperature increases in the Arctic region, contaminants currently locked in the ice and soils will be released into the air, water, and land. This trend, along with increased human activity within and below the Arctic, will result in increased introduction of contaminants into the Arctic, including both persistent pollutants (e.g., persistent organic pollutants and mercury) and airborne pollutants (e.g., soot).
6. Implementation: In carrying out this policy as it relates to environmental protection and conservation of natural resources, the Secretaries of State, the Interior, Commerce, and Homeland Security and the Administrator of the Environmental Protection Agency, in coordination with heads of other relevant executive departments and agencies, shall:
   a. In cooperation with other nations, respond effectively to increased pollutants and other environmental challenges;
   b. Continue to identify ways to conserve, protect, and sustainably manage Arctic species and ensure adequate enforcement presence to safeguard living marine resources, taking account of the changing ranges or distribution of some species in the Arctic. For species whose range includes areas both within and beyond United States jurisdiction, the United States shall

continue to collaborate with other governments to ensure effective conservation and management;

c. Seek to develop ways to address changing and expanding commercial fisheries in the Arctic, including through consideration of international agreements or organizations to govern future Arctic fisheries;
d. Pursue marine ecosystem-based management in the Arctic; and
e. Intensify efforts to develop scientific information on the adverse effects of pollutants on human health and the environment and work with other nations to reduce the introduction of key pollutants into the Arctic.

IV. Resources and Assets

A. Implementing a number of the policy elements directed above will require appropriate resources and assets. These elements shall be implemented consistent with applicable law and authorities of agencies, or heads of agencies, vested by law, and subject to the availability of appropriations. The heads of executive departments and agencies with responsibilities relating to the Arctic region shall work to identify future budget, administrative, personnel, or legislative proposal requirements to implement the elements of this directive.

[1] These policies and authorities include Freedom of Navigation (PDD/NSC-32), the U.S. Policy on Protecting the Ocean Environment (PDD/NSC-36), Maritime Security Policy (NSPD-41/HSPD-13), and the National Strategy for Maritime Security (NSMS).[291]

End Notes

1 Except for the subsection on the Arctic and the U.N. Convention on the Law of the Sea, this section was prepared by Ronald O'Rourke, Specialist in Naval Affairs, Foreign Affairs, Defense, and Trade Division.

2 On November 25, 2008, voters in Greenland approved a referendum for greater autonomy that some observers view as a step toward eventual independence from Denmark. (Alan Cowell, "Greenland Vote Favors Independence," *New York Times*, November 26, 2008.)

3 For more on the Arctic Council on the Internet, see http://www.arctic-council.org/.

4 Title II of P.L. 98-373 is the National Critical Materials Act of 1984.

5 As codified, the definition reads, "As used in this chapter.... "

6 For discussions and (in some cases) maps, see Susan Joy Hassol, *Impacts of a Warming Arctic* [Executive Summary]. Cambridge, Cambridge University Press, 2004, p. 4, available at

http://amap.no/acia/; Oran R. Yong and Niels Einarsson, *Arctic Human Development Report*, Stefansson Arctic Institute, Akureyri, Iceland, 2004, pp. 17-18, available at http://www.svs.is/AHDR/AHDR%20chapters/English%20version/Chapters%20PDF.htm; and Hugo Ahlenius, editor in chief et al. *Vital Arctic Graphics, People and Global Heritage on Our Last Wild Shores*, UNEP/GRID-Arendal, Arendal Norway, p. 6, available at http://www.grida.no/publications/vg/arctic/.

[7] A map showing the line that results from 10° isotherm definition is available at https://www.cia.gov/library/publications/the-world-factbook/reference_maps/pdf/arctic.pdf.

[8] Discussion entitled "Geographical Coverage," available at http://www.amap.no/ (click on "About AMAP" and then the tab "Geographical coverage.")

[9] Discussion entitled "Geographical Coverage," available at http://www.amap.no/ (click on "About AMAP" and then the tab "Geographical coverage.")

[10] Title II of P.L. 98-373 is the National Critical Materials Act of 1984.

[11] These words are taken from the official title of P.L. 98-373. (Arctic Research and Policy Act of 1984 is the short title of Title I of P.L. 98-373.) The remainder of P.L. 98-373's official title relates to Title II of the act, the short title of which is the National Critical Materials Act of 1984.)

[12] The IARPC currently includes more than a dozen federal agencies, departments, and offices. Additional information on the IARPC is available at http://www.nsf.gov/od/opp/arctic/iarpc/start.jsp.

[13] NSF used to conduct its Arctic research activities through a Division of Arctic Sciences (ARC) that formed part of an Office of Polar Programs (OPP). NSF states that in FY2013, a realignment of offices previously under the Office of the Director was implemented, and that as a part of this realignment, OPP became PLR within GEO. (National Science Foundation, *FY 2014 Budget Request to Congress*, April 10, 2013, Footnote 2 in Summary Tables—3 (pdf page 25 of 506). The document is available at http://www.nsf.gov/about/budget/fy2014/pdf/EntireDocument_fy2014.pdf.)

[14] National Science Foundation, *FY 2014 Budget Request to Congress*, April 10, 2013, p. GEO-15 (pdf page 127 of 506).

[15] National Science Foundation, *FY 2014 Budget Request to Congress*, April 10, 2013, p. GEO-1 (pdf page 113 of 506).

[16] National Science Foundation, *FY 2014 Budget Request to Congress*, April 10, 2013, pp. GEO-15 and GEO-16 (pdf pages 127 and 128 of 506).

[17] National Science Foundation, *FY 2014 Budget Request to Congress*, April 10, 2013, p. GEO-16 (pdf page 128 of 506).

[18] National Science Foundation, *FY 2014 Budget Request to Congress*, April 10, 2013, p. GEO-17 (pdf page 129 of 506).

[19] CRS communication with State Department official, October 8, 2010.

[20] *National Security Strategy*, Washington, May 2010, p. 50. The quoted sentence constitutes the entirety of the document's comments specifically on the Arctic. It is the final sentence of a section on "sustain[ing] broad cooperation on key global challenges" that includes longer discussions on climate change, peacekeeping and armed conflict, pandemics and infectious disease, transnational criminal threats and threats to governance, and safeguarding the global commons.

[21] *National Strategy for the Arctic Region*, May 2013, 11 pp.; accessed May 14, 2013, at http://www.whitehouse.gov/ sites/default/files/docs/nat_arctic_strategy.pdf. The document includes a cover letter from President Obama dated May 10, 2013.

[22] *National Strategy for the Arctic Region* states on page 6 that the "lines of effort" it describes are to be undertaken "[t]o meet the challenges and opportunities in the Arctic region, and in furtherance of established Arctic Region Policy," at which point there is a footnote referencing the January 2009 Arctic policy directive.

[23] A footnote in the document at this point states: "Traditional knowledge refers to a body of evolving practical knowledge based on observations and personal experience of indigenous

communities over an extensive, multigenerational time period. (BOEM Ocean Science, Vol. 9, Issue 2, May/April/June 2012, page 4)."

[24] *National Strategy for the Arctic Region*, May 2013, pp. 2-3.

[25] This section prepared by Marjorie Ann Browne, Specialist in International Relations, Foreign Affairs, Defense, and Trade Division.

[26] The United States is party to the four conventions adopted in 1958: Convention on the Territorial Sea and the Contiguous Zone, Convention on the High Seas, Convention on the Continental Shelf, and Convention on Fishing and Conservation of the Living Resources of the High Seas.

[27] Treaty Document 103-39.

[28] Other relevant provisions of the Convention, applicable depending on the extent of Arctic melting, relate to navigation, high seas freedoms, fisheries, and exclusive economic zones.

[29] The continental shelf is the under-sea extension of a coastal state's land territory. Article 76 of the Convention defines the continental shelf, *inter alia*, as "the seabed and subsoil of the submarine areas that extend beyond its [coastal state's] territorial sea throughout the natural prolongation of its land territory to the outer edge of the continental margin."

[30] A coastal State party has 10 years from the entry into force of the Convention for submission of information on its proposed limits. In May 2001, the Meeting of States Parties to the Convention decided that for any State for which the Convention entered into force before May 13, 1999, the date of commencement of the 10-year time period for making submissions to the Commission is May 13, 1999.

[31] Annex II, Article 9. Article 83 of the Convention provides that questions relating to these boundary delimitation disputes shall be resolved by agreement between the States or by the Dispute Settlement options set forth in Part XV of the Convention.

[32] Article 76, para. 8.

[33] For more information, see http://www.state.gov/g/oes/continentalshelf/index.htm.

[34] This section prepared by Jane Leggett, Specialist in Energy and Environmental Policy, Resources, Science, and Industry Division.

[35] In scientific analyses, "ice-free" does not necessarily mean "no ice." The definition of "ice-free" or sea ice "extent" or "area" varies across studies. Sea ice "extent" is one common measure, equal to the sum of the area of grid cells that have ice concentration of less than a set percentage—frequently 15%. For more information, see the National Snow and Ice Data Center, http://nsidc.org/seaice/data/terminology.html.

[36] Muyin Wang and James E. Overland, "A Sea Ice Free Summer Arctic within 30 Years?," *Geophysical Research Letters* 36, no. L07502 (April 3, 2009): 10.1029/2009GL037820; Marika Holland, Cecilia M. Bitz, and Bruno Tremblay, "Future abrupt reductions in the summer Arctic sea ice," *Geophysical Research Letters* 33, no. L23503 (2006); But see also Julien Boé, Alex Hall, and Xin Qu, "Sources of spread in simulations of Arctic sea ice loss over the twenty-first century," *Climatic Change* 99, no. 3 (April 1, 2010): 637-645; I. Eisenman and J. S. Wettlaufer, "Nonlinear threshold behavior during the loss of Arctic sea ice," *Proceedings of the National Academy of Sciences* 106, no. 1 (January 6, 2009): 28-32; Dirk Notz, "The Future of Ice Sheets and Sea Ice: Between Reversible Retreat and Unstoppable Loss," *Proceedings of the National Academy of Sciences* 106, no. 49 (December 8, 2009): 20590-20595.

[37] Overland et al. state that "a warm Arctic-cold continent pattern represents a paradox of recent global warming: there is not a uniform pattern of temperature increases" due to a set of newly recognized processes described in Overland, J. E, K. R Wood, and M. Wang. "Warm Arctic-cold Continents: Climate Impacts of the Newly Open Arctic Sea." *Polar Research* 30 (2011). The authors raise a critical, unanswered question, "Is the observed severe mid-latitude weather in two adjacent years simply due to an extreme in chaotic processes alone, or do they included a partial but important Arctic forcing and connection due to recent changing conditions?" In other words, are recent patterns random anomalies, or might we expect more of the same?; among other examples, see also Lim, Young-Kwon, and

Siegfried D. Schubert. "The Impact of ENSO and the Arctic Oscillation on Winter Temperature Extremes in the Southeast United States." *Geophysical Research Letters* 38, no. 15 (August 11, 2011): L15706.

[38] There was a regionally warm period in the Arctic from the mid-1920s to around 1940, which scientists have assessed to have been driven by natural climate variability. They have found that period to be distinctly different from the recent multi-decadal warming, in part because the early 20th century warming was concentrated in the northern high latitudes. See, for example, Figure 2, upper left graphic, in Geophysical Fluid Dynamics Laboratory, "Simulatoin of Early 20th Century Warming," at http://www.gfdl.noaa.gov/early-20th-century-global-warming.

[39] Steele, Michael, Wendy Ermold, and Jinlun Zhang. "Arctic Ocean Surface Warming Trends over the Past 100 Years." *Geophysical Research Letters* 35, no. 2 (January 29, 2008): L02614.

[40] J. Overland, J. Walsh, and M. Wang, *Arctic Report Card - Atmosphere* (NOAA Arctic Research Program, October 6, 2008), http://www.arctic.noaa.gov/reportcard/atmosphere.html.

[41] National Snow and Ice Data Center, "Arctic sea ice extent breaks 2007 record low" (August 27, 2012); Japanese Aerospace Exploration Agency, "A new record minimum of the Arctic sea ice extent was set on 24 August 2012"; Arctic ROOS (Norway), "Daily Updated Time series of Arctic sea ice area and extent derived from SSMI data provided by NERSC," at http://arctic-roos.org/observations/satellite-data/sea-ice/ice-area-and-extent-in-arctic.

[42] William L. Chapman and John E. Walsh, "Simulations of Arctic Temperature and Pressure by Global Coupled Models," *Journal of Climate* 20, no. 4 (February 1, 2007): 609-632.

[43] See footnote 35. Also, although one Canadian scientist has predicted that recurrent ice-free summers may begin sometime between 2013 and 2020, this is not consistent with other climate models' projections.

[44] This section prepared by Carl Ek, Specialist in International Relations, Foreign Affairs, Defense, and Trade Division.

[45] "Russia, Canada Make Competing Claims To Arctic Resources," *The Canadian Press*, September 16, 2010.

[46] "5 Countries Agree To Talk, Not Compete, Over the Arctic," *New York Times*, May 29, 2008.

[47] For additional information, see Natalie Mychajlyszyn, *The Arctic: Geopolitical Issues*, Canadian Library of Parliament, October 24, 2008.

[48] "Russia, Norway Sign Deal On Barents Sea Border, Seek More Development in Mineral-rich Arctic," *Associated Press*, September 15, 2010.

[49] Cannon quoted in "Canada Seeks To Settle Arctic Borders," *Agence France Presse*, August 20, 2010. For additional information concerning Canada's August statement on Arctic policy, see "Statement on Canada's Arctic foreign policy: Exercising sovereignty and promoting Canada's Northern Strategy abroad," Foreign Affairs and International Trade Canada website, http://www.international.gc.ca/polar-polaire/canada_arctic_foreign_policy-la_politique_etrangere_du_canada_pour_arctique.aspx?lang=eng.

[50] "Canada and Denmark Sign Arctic Cooperation Arrangement," *Targeted News Service*, May 17, 2010.

[51] This section prepared by John Frittelli, Specialist in Transportation Policy, Resources, Science, and Industry Division.

[52] Extended daylight hours in the Arctic during the summer may also be an advantage.

[53] Presentation by Stephen Carmel, Senior Vice President, Maersk Line Ltd., Halifax International Security Forum, *Arctic Security: The New Great Game?* November 21, 2009, available at http://fora.tv/.

[54] A third but more remote possibility is a route directly over the North Pole.

[55] In the summer of 2009, two German-owned ships are believed to be the first non-Russian ships to transit the NSR, sailing from Korea to deliver oil-service equipment to the Russian Arctic and then sailing on to Rotterdam.

[56] This was the route pioneered by the *SS Manhattan*, an oil tanker modified for ice breaking in 1969 to carry Alaskan North Slope oil to the Atlantic. This was the first commercial passage through the NWP, but the building of the Alaskan pipeline was found to be the more economical means of transporting oil from the North Slope to the lower 48 states.

[57] Although the NWP is often compared to the alternative route through the Panama Canal in terms of distance and sailing days from Asia to the U.S. east coast, another alternative to consider is the shorter and faster transcontinental rail route across Canada or the United States. The Panama Canal is undergoing an expansion project, expected to be completed by 2015, to allow larger ships with about three times the cargo capacity to pass through.

[58] ITAR-TASS, "Rosatomflot Ends Shipping Season Along Northern Sea Route," November 29, 2011.

[59] "U.S. Needs to Prepare for Arctic Traffic Surge," *Anchorage Daily News*, February 12, 2008, p. A4.

[60] "Northern Waters," *The Globe and Mail* (Canada), August 14, 2008, p. A8.

[61] See http://www.aeco.no/index.htm for more information.

[62] S.E.L. Howell and J.J. Yackel, "A Vessel Transit Assessment of Sea Ice Variability in the Western Arctic, 1969- 2002: Implications for Ship Navigation," *Canadian Journal of Remote Sensing*, vol. 30, no. 2, 2004.

[63] "Arctic Unlikely to See Major Shipping Growth," *New Zealand Transport and Logistics Business Week*, April 24, 2008.

[64] Presentation by Stephen Carmel, Senior Vice President, Maersk Line Ltd., Halifax International Security Forum, *Arctic Security: The New Great Game?* November 21, 2009, available at http://fora.tv/.

[65] In July and August 2010, NOAA surveyed the Bering Straits area in order to update its charts but stated that it will take more than 25 years to map the prioritized areas of navigational significance in U.S. Arctic waters. See http://www.noaanews.noaa.gov/stories2010/20100720_fairweather.html.

[66] Ice reporting that currently exists is intended for scientists not mariners.

[67] For further information, see http://www.poa.usace.army.mil/en/cw/AKPortsStudy.htm, and FY2013 USACE Budget Justification, p. POD-5.

[68] See http://www.cmts.gov/Activities/ActionTeams.aspx.

[69] See http://www.imo.org/ for more information.

[70] See MSC/Circ. 1056/MEPC/Circ.399, at http://www.imo.org/.

[71] Edwin H. Anderson, "Polar Shipping, The Forthcoming Polar Code and Implications for the Polar Environments," *Journal of Maritime Law and Commerce,* v. 43, no. 1, January 2012.

[72] The Coast Guard is studying shipping routes through the Bering Strait for possible safety enhancements. See 75 FR 68568, November 8, 2010.

[73] This section prepared by Curry Hagerty, Specialist in Energy and Natural Resources Policy, Resources, Science, and Industry Division; Peter Folger, Specialist in Energy and Natural Resources Policy, Resources, Science, and Industry Division; and Marc Humphries, Analyst in Energy Policy, Resources, Science, and Industry Division.

[74] Data provided by written communication to CRS from Bureau of Safety and Environmental Enforcement on October 4-11, 2012. There are no federally regulated production facilities or pipelines in the Alaska region.

[75] One way of delineating the extent of the continental margins that is currently under way in the region is by the terms of Article 76 of the U.N. Convention on the Law of the Sea (UNCLOS).

[76] CRS Report R41132, *Outer Continental Shelf Moratoria on Oil and Gas Development*, by Curry L. Hagerty.

[77] FY2004 DOI Appropriations (P.L. 108-108). Furthermore, the Continuing Appropriations Resolution 2009 (P.L. 110-329) did not extend the annual congressional moratorium on oil and gas leasing activities in the lower 48 states. On March 11, 2009, the Omnibus Appropriations Act, 2009 (P.L. 111-8) was enacted without moratorium provisions,

confirming that the oil and gas development bans in federal waters along the Atlantic and Pacific coasts, parts of Alaska, and the Gulf of Mexico that had been in place since 1982 had not been restored in 2009 appropriations measures.

[78] "*Withdrawal of Certain Areas of the United States Outer Continental Shelf from Leasing Disposition*," §12(a) OCSLA, 43 U.S.C. 1341(a) (March 31, 2010). This memorandum is available at http://www.doi.gov/whatwedo/energy/ ocs/upload/2010alaska-mem-rel.pdf.

[79] See statement of the Alaska OCS Region at http://www.boem.gov/About-BOEM/BOEM-Regions/Alaska-Region/ Index.aspx. On March 27, 2012, BOEM issued a Request for Interest, with respect to the Cook Inlet planning area.

[80] For example with respect to pre-leasing activities in the Cook Inlet, on March 27, 2012, BOEM issued a Request for Interest, with respect to the Cook Inlet planning area. In light of responses to the Request, BOEM decided to proceed with the pre-sale process for the Cook Inlet and to place the date for a potential lease sale in 2016 to allow time to complete the necessary steps to develop additional resource and environmental information, and conduct analysis required by NEPA.

[81] For statements of legislative interest, see House Committee on Natural Resources 113th Congress Oversight Plan (adopted January 23, 2013). For statements of interest by President Obama, see 2013 State of the Union address at http://www.whitehouse.gov/state-of-the-union-2013.

[82] 43 U.S.C. §1344. The current Five Year Program took effect August 27, 2012, and is expected to remain in effect until 2017. 77 *Federal Register* 40080 (July 6, 2012).

[83] On April 20, 2010, an explosion occurred at the *Deepwater Horizon* drilling platform in the Gulf of Mexico, resulting in 11 fatalities. The platform sank into the Gulf on April 22, 2010, amid a significant release of oil at the sea floor. According to the National Incident Command's Flow Rate Technical Group estimate, the well released approximately 206 million gallons (4.9 million barrels) of oil before it was contained July 15, 2010.

[84] Shell is a global group of energy and petrochemical companies with 90,000 employees in more than 80 countries. In the United States, Shell operates in 50 states as an oil and natural gas producer, gasoline and natural gas marketer, and petrochemical manufacturer. See more at http://www.shell.com.

[85] Among other approvals, on August 4, 2011, BOEM issued a conditional approval of Shell's revised exploration plan (EP) for the Beaufort Sea to drill up to four exploration wells starting in summer 2012. On December 16, 2011, BOEM issued conditional approval of Shell's revised EP in the Chukchi Sea to drill up to six exploration wells beginning in the 2012 drilling season.

[86] Shell's *Kulluk* ran aground off the southern coast of Alaska in a storm. The rig was towed to safe harbor, with no serious injuries to the crew.

[87] The U.S. Coast Guard conducted this review with the Bureau of Safety and Environmental Enforcement (BSEE) and the National Transportation Safety Board providing technical assistance. For more details on this review see "*Secretary Salazar Launches Expedited Assessment of 2012 Arctic Operations High-Level, Expedited Review of the 2012 Offshore Drilling Program in the Beaufort and Chukchi Seas Will Assess Activities, Provide Lessons Learned*," (January 8, 2013), at http://www.doi.gov/news/pressreleases/secretary-salazar-launches-expedited-assessment-of-2012-arctic- operations.cfm.

[88] For Mr. Luthi's full statement see *Keep Cool Head on Arctic Energy Development*, by Amy Harder, National Journal, reporting comments submitted by Randall Luthi, President of the National Ocean Industries Association (January 24, 2013),available at http://energy.nationaljournal.com/2013/01/are-arctic-oildrilling-challen.php.

[89] For a full discussion of this perspective see *We're Not Yet Arctic-Ready* by Marilyn Heiman, Director of the Pew Environment Group's U.S. Arctic Program (January 21, 2013). This statement is available at http://energy.nationaljournal.com/2013/01/are-arctic-oildrilling-challen.php.

[90] Prior to DOI approvals issued in 2012, Shell had attempted to establish an exploratory drilling program offshore Alaska in the Beaufort and Chukchi Seas for several years. Shell's long-standing goal has been to confirm a major discovery of oil in commercially-viable quantities in the Alaskan Arctic Ocean. For a comprehensive discussion of case law featuring Shell as a defendant and as a plaintiff in litigation over federal leasing in the Alaska program areas see CRS Report WSLG332, *A Preemptive Legal Strike? Shell Seeks Injunction to Prevent Environmental Group Challenges to Drilling Permits*, by Adam Vann.

[91] DOI required Shell to include in its drilling plan an in-theater subsea containment system capable of functioning in the event of an emergency and subject to review and oversight by federal agencies responsible for Arctic exploration. For further information see "*Secretary Salazar Launches Expedited Assessment of 2012 Arctic Operations*," January 8, 2013, available at http://www.doi.gov/news/pressreleases/secretary-salazar-launches-expeditedassessment- of-2012- arctic-operations.cfm.

[92] See Report to the Secretary of the Interior; Review of Shell's 2012 Alaska Offshore Oil and Gas Exploration Program (March 8, 2013), at http://www.doi.gov/news/pressreleases/upload/Shell-report-3-8-13-Final.pdf.

[93] A seismic array is typically a long string or streamer of geophones—acoustic devices used for recording seismic signals—towed behind a ship while the ship traverses a prospective oil and gas-bearing portion of the seafloor. The seismic signals are processed and interpreted to give a cross-section or three-dimensional image of the subsurface.

[94] See USGS Circum-Arctic Resource Appraisal website at http://energy.usgs.gov/arctic/.

[95] USGS Fact Sheet 2008-3049: Circum-Arctic Resource Appraisal: Estimates of Undiscovered Oil and Gas North of the Arctic Circle, http://pubs.usgs.gov/fs/2008/3049/.

[96] Donald L. Gautier et al., "Assessment of Undiscovered Oil and Gas in the Arctic," *Science*, vol. 324, no. 5931, pp. 1175-1179 (2009).

[97] The Associated Press, "Greenland Opens to Oil Firms; Melting Ice Unlocks Reserves," International Herald Tribune (January 14, 2008), at http://www.iht.com/articles/ap/2008/01/14/business/NA-FIN-US-Greenland-Oil-Companies.php.

[98] *Secretary Salazar Unveils Arctic Studies Initiative that will Inform Oil and Gas Decisions for Beaufort and ChukchiSeas*: Press Release April 13, 2010, http://www.doi.gov/news/pressreleases/2010_04_13_releaseA.cfm.

[99] U.S. Geological Survey Fact Sheet 2011-3048, 4 p. (2011).

[100] Tony Halpin, "President Medvedev Threatens Russian Arctic Annexation," Times Online (September 18, 2008), at http://www.timesonline.co.uk/tol/news/world/europe/article 4773567.ece.

[101] The purpose of the U.S. Extended Continental Shelf Project is to establish the full extent of the continental shelf of the United States, consistent with international law. Involved in this mission are the U.S. Coast Guard (USCG), Department of State (DOS), and the University of New Hampshire (UNH). NOAA has the lead in collecting bathymetric data. The U.S. Geological Survey has the lead in collecting seismic data.

[102] See USGS, U.S. Extended Continental Shelf Project, at http://www.state.gov/g/oes/continentalshelf/.

[103] A prominent directive issued by the Bush Administration addressed, among other issues, national security and maritime transportation. National Security Presidential Directive/NSPD-66 and Homeland Security Presidential Directive/HSPD-25, at http://www.whitehouse.gov/news/releases/2009/01/20090112-3.html.

[104] This initiative is part of an effort Arctic Research Commission Chairman Fran Ulmer began in December 2011. For a full discussion of this initiative see "*Obama Administration Outlines Arctic Energy Policy Initiatives*," Oil and Gas Journal (February 7, 2012), available at http://www.ogj.com/articles/2012/02/obama-administration-outlines-arctic- energy-policy-initiatives.html.

[105] Holland-Bartels, Leslie, and Pierce, Brenda, eds., 2011, An evaluation of the science needs to inform decisions on Outer Continental Shelf energy development in the Chukchi and Beaufort Seas, Alaska: U.S. Geological Survey Circular 1370, 278 p. (2011). This report is available at http://pubs.usgs.gov/circ/1370/.

[106] Don Gautier, Research Geologist, USGS, quoted in "Countries in Tug-of-War Over Arctic Resources," CNN.com (January 2, 2009), at http://www.cnn.com/2009/TECH/science/01/02/arctic.rights.dispute/index.html?iref=newssearch.

[107] Carolyn Fitzpatrick, "Heavy Haul in the High North," Railway Gazette International (July 24, 2008), at http://www.railwaygazette.com/news/single-view/view/10/heavy-haul-in-the-high-north.html.

[108] Cecelia Jamasmie, *Melting Arctic is nothing but good news for the mining industry: Lloyd's/Chatham House*, Mining.com, April 13, 2012, http://www.mining.com/melting-arctic-is-nothing-but-good-news-for-the-mining- industry-lloydschatam-house/.

[109] Charles Emmerson and Glada Lahn, *Arctic Opening: Opportunity and Risk in the High North*, Chatham House, 2012, http://www.lloyds.com/~/media/Files/News%20and%20Insight/360%20Risk%20Insight/ Arctic_Risk_Report_20120412.pdf.

[110] See National Oceanic and Atmospheric Administration, Arctic Change, at http://www.arctic.noaa.gov/detect/land- road.shtml?page=land.

[111] This section prepared by Jonathan L. Ramseur, Specialist in Environmental Policy, Resources, Science, and Industry Division.

[112] Arctic Monitoring and Assessment Programme (AMAP), Arctic Oil and Gas 2007 (2008).

[113] Arctic Monitoring and Assessment Programme (AMAP), Arctic Oil and Gas 2007 (2008).

[114] See this report's section "Implications for Sea Transportation," by John Fritelli.

[115] For example, the *Exxon Valdez* spilled approximately 11 million gallons of oil, but its carrying capacity was approximately 60 million gallons.

[116] See this report's section "Implication of Changes in the Arctic for Oil, Gas, and Mineral Exploration and Development," by Peter Folger and Marc Humphries.

[117] Larger oil spills occurred during the 1991 Iraq War, but many of those spills were deliberate. A 1910-1911 onshore oil blowout in the California San Joaquin Valley is reported to have spilled 9.4 million barrels of crude oil (almost 400 million gallons).

[118] See CRS Report R41531, *Deepwater Horizon Oil Spill: The Fate of the Oil*, by Jonathan L. Ramseur.

[119] National Research Council (NRC) of the National Academies of Science, *Oil in the Sea III: Inputs, Fates, and Effects* (2003).

[120] See CRS Report RL33705, *Oil Spills in U.S. Coastal Waters: Background and Governance*, by Jonathan L. Ramseur; and Dagmar Etkin (Environmental Research Consulting), Analysis of U.S. Oil Spillage, Prepared for American Petroleum Institute, August 2009.

[121] National Research Council of the National Academies of Science, *Cumulative Environmental Effects of Oil and Gas Activities on Alaska's North Slope* (2003).

[122] See, for example, Minerals Management Service (MMS), Outer Continental Shelf Oil & Gas Leasing Program: 2007-2012, Final Environmental Impact Statement, April 2007, Chapter 4; MMS, Proposed Gulf of Mexico OCS Oil and Gas Lease Sale 206, Central Planning Area, Environmental Assessment, October 2007;

[123] Letter from Marvin E. Odum, President, Shell Oil Company to S. Elizabeth Birnbaum, Minerals Management Service (May 14, 2010). Cited in a staff paper from the National Commission on the BP Deepwater Horizon Oil Spill and Offshore Drilling ("The Challenges of Oil Spill Response in the Arctic," January 2011).

[124] AMAP, *Arctic Oil and Gas 2007 (2008).*

[125] AMAP, *Arctic Oil and Gas 2007* (2008).

[126] World Wildlife Fund, *Oil Spill: Response Challenges in Arctic Waters* (2007).

[127] Coastal Response Research Center, *Opening the Arctic Seas: Envisioning Disasters and Framing Solutions* (2009), partnership between the National Oceanic and Atmospheric Administration and the University of New Hampshire.

[128] Although the response gap in the Arctic has not been quantified, a recent estimate of Prince William Sound (PWS) may be instructive. A 2007 study found a response gap for PWS of 38% for the time of the study period (65% during the winter season). Note that PWS has existing infrastructure for response, while the more remote Arctic areas do not. Nuka Research and Planning Group, LLC, *Response Gap Estimate for Two Operating Areas in Prince William Sound, Alaska* (2007), Report to Prince William Sound Regional Citizens' Advisory Council.

[129] Arthur Brooks (Commander, 17th Coast Guard District) Coast Guard Journal, Arctic Journal (April 7, 2008).

[130] U.S. Coast Guard, *Report to Congress: U.S. Coast Guard Polar Operations*, December 2008.

[131] A single road connects Deadhorse, Alaska and the Prudhoe Bay with central Alaska (Fairbanks). For more details and maps of the area, see Nuka Research and Planning Group, *Oil Spill Prevention and Response in the U.S. Arctic Ocean: Unexamined Risks, Unacceptable Consequences*, Commissioned by Pew Environment Group, November 2010.

[132] Government Accountability Office, *Coast Guard: Efforts to Identify Arctic Requirements Are Ongoing, but More Communication about Agency Planning Efforts Would Be Beneficial*, GAO-10-870, September 2010.

[133] Transportation Research Board of the National Academy of Sciences, *Risk of Vessel Accidents and Spills in the Aleutian Islands: Designing a Comprehensive Risk Assessment* (2008), Special Report 293, National Academies Press. Washington, DC.

[134] Coastal Response Research Center, *Opening the Arctic Seas: Envisioning Disasters and Framing Solutions*, (2009), partnership between the National Oceanic and Atmospheric Administration and the University of New Hampshire. See also, U.S. Arctic Research Commission, (Draft) *White Paper: U.S. Arctic Research Commission Recommends Steps to Expanded U.S. Funding for Arctic/Subarctic Oil Spill Research* (February 24, 2010).

[135] Arctic Monitoring and Assessment Programme (AMAP), *Arctic Oil and Gas 2007* (2008).

[136] National Research Council, *Cumulative Environmental Effects of Oil and Gas Activities on Alaska's North Slope* (2003).

[137] Arturo A. Keller and Kristin Clark, *Oil Recovery with Novel Skimmer Surfaces under Cold Climate Conditions* (2007), prepared for the Minerals Management Service.

[138] World Wildlife Fund, *Oil Spill: Response Challenges in Arctic Waters* (2007). For further discussion of issues relating to oil spills, see CRS Report RL33705, *Oil Spills in U.S. Coastal Waters: Background and Governance*, by Jonathan L. Ramseur.

[139] Coastal Response Research Center, *Opening the Arctic Seas: Envisioning Disasters and Framing Solutions*, (2009), partnership between the National Oceanic and Atmospheric Administration and the University of New Hampshire.

[140] This section prepared by Eugene Buck, Specialist in Natural Resources Policy, Resources, Science, and Industry Division.

[141] Erik J. Molenaar and Robert Corell, *Arctic Fisheries*, Arctic Transform, February 9, 2009; available at http://arctic- transform.org/download/FishBP.pdf.

[142] 74 *Federal Register* 56734-56746, November 3, 2009.

[143] Information on the symposium is available at http://www.nprb.org/iafs2009/.

[144] The website is available at http://alaskafisheries.noaa.gov/sustainablefisheries/arctic/.

[145] Prepared by Lynne Corn and Eugene Buck, Specialists in Natural Resources Policy, Resources, Science, and Industry Division.

[146] Fish and Wildlife Service, "Endangered and Threatened Wildlife and Plants; Review of Native Species That Are Candidates for Listing as Endangered or Threatened; Annual Notice of Findings on Resubmitted Petitions; Annual Description of Progress on Listing Actions; Proposed Rule," 76 *Federal Register* 66370-66439, October 26, 2001.

[147] 73 *Federal Register* 79822-79828.

[148] 75 *Federal Register* 65239-65248.

[149] 74 *Federal Register* 53683-58696, October 20, 2009.

[150] 75 *Federal Register* 77476-77495.

[151] 75 *Federal Register* 77496-77515.

[152] Among biologists, it is traditionally said that a species faced with extreme change can respond in three basic ways: "migrate, mutate, or die." When change is rapid enough, mutation (accompanied by natural selection of individuals within the population more suited to the changed environment) may not be able to occur fast enough, leaving migration and death as the only options. The problem of response rate is more severe for species that reproduce slowly (e.g., polar bears) and less severe for species that reproduce rapidly (e.g., algae).

[153] The efficacy and the effect of this tactic is often questioned, since natural migration is unlikely to involve the entire suite of species in an ecosystem (e.g., host plants might not move north (or up) as fast as their moth herbivores, nor as fast as the birds that depend on the moths). Moreover, the southerners will not find a land of sterile bare dirt—the species that are already there may be threatened themselves by the competition from the new arrivals, perhaps tipping the balance and pushing still more species toward extinction.

[154] Under the MMPA, both NMFS and FWS have responsibility for additional marine mammal species (e.g., manatees, sea otters, dolphins) which are not currently found in the Arctic.

[155] This section was originally prepared by Roger Walke, who was a Specialist in American Indian Policy, Domestic Social Policy Division, until his retirement from CRS in October 2010. It has been updated by Jane A. Leggett, Specialist in Environmental and Energy Policy in CRS's Resources, Science and Industry Division.

[156] *Arctic Human Development Report*, ed. Joan Nymand Larsen et al. (Akureyri, Iceland: Stefansson Arctic Institute, 2004), p. 47; this report is subsequently cited in this section as *AHDR.* The seven countries are Canada, Denmark-Greenland, Finland, Norway, Russia, Sweden, and the United States.

[157] John F. Hoffecker, *A Prehistory of the North: Human Settlement of the Higher Latitudes* (New Brunswick, NJ: Rutgers University Press, 2005), pp. 8, 81, 112-115.

[158] *AHDR*, pp. 47, 53; David Crystal, *Cambridge Encyclopedia of Language*, 2nd ed. (Cambridge: Cambridge University Press, 1997), chap. 50; *Ethnologue: Languages of the World*, 16th ed., ed. M. Paul Lewis (Dallas: SIL International, 2009), available at http://www.ethnologue.com/. The number of languages and language families varies not only with definitions of the Arctic but with definitions of languages and language families.

[159] Jim Berner et al., *Arctic Climate Impact Assessment* (Cambridge: Cambridge University Press, 2005), chapter 12; this report is subsequently cited in this section as *ACIA*.

[160] *ACIA*, pp. 654-655.

[161] Fennoscandia refers to the Scandinavian Peninsula, Finland, the Kola Peninsula of Russia, and certain parts of Russia bordering on Finland.

[162] Janet Martin, *Treasure in the Land of Darkness: The Fur Trade and Its Significance for Medieval Russia* (Cambridge: Cambridge University Press, 1986), pp. 41-42; James Forsyth, *A History of the Peoples of Siberia: Russia's North Asian Colony, 1581-1990* (Cambridge: Cambridge University Press, 1992), pp. 69-83, 102; Lassi K. Heininen, "Different Images of the Arctic and the Circumpolar North in World Politics," in *Knowledge and Power in the Arctic*, Proceedings at a Conference in Rovaniemi, April 16-18, 2007, Arctic Centre Reports 48, ed. Paula Kankaanpaa et al. (Rovaniemi, Finland: University of Lapland, Arctic Centre, 2007), p. 125.

[163] James W. VanStone, "Exploration and Contact History of Western Alaska," and David Damas, "Copper Eskimo," and J. Garth Taylor, "Historical Ethnography of the Labrador Coast," in *Handbook of North American Indian: Vol. 5, Arctic,* vol. ed. David Damas, gen. ed. William C. Sturtevant (Washington: Smithsonian, 1984), pp. 149-155, 408, 509- 510.

[164] Inge Kleivan, "History of Norse Greenland," in *Handbook, Vol. 5, Arctic, op. cit.,*, pp. 549-555; Finn Gad, "Danish Greenland Policies," in *Handbook of North American Indians: Vol. 4, History of Indian-White Relations*, vol. ed. Wilcomb E. Washburn, gen. ed. William C. Sturtevant (Washington: Smithsonian, 1988), p. 110.

[165] Parkinson, Alan J. The Arctic Human Health Initiative. Washington DC: U.S. Department of Health and Human Services, 2006.

[166] *AHDR*, pp. 19, 29. Estimates of Arctic indigenous populations are complicated by varying definitions not only of the Arctic but also of indigenous peoples; for instance, Russia does not count some non-European Arctic ethnic groups, such as the Yakut, as "indigenous minorities" (see "Peoples of the Arctic: Characteristics of Human Populations Relevant to Pollution Issues," in *AMAP Assessment Report: Arctic Pollution Issues*, ed. Simon J. Wilson et al. (Oslo: Arctic Monitoring and Assessment Programme, 1998), pp. 167-169; this report is subsequently cited in this section as *AMAP 1998*.

[167] *AMAP 1998*, chapter 5; see also Birger Poppel et al., *SLiCA Results*, Survey of Living Conditions in the Arctic (Anchorage: Institute of Social and Economic Research, University of Alaska Anchorage, 2007), pp. 4-7, http://www.arcticlivingconditions.org.

[168] Annika E. Nilson and Henry P. Huntington, *Arctic Pollution 2009* (Oslo: Arctic Monitoring and Assessment Programme, 2009), p. 39-41; this report is subsequently cited in this section as *AMAP 2009*.

[169] *ACIA*, p. 1000.

[170] *SLiCA Results, op.cit.*, pp. v, 4-8.

[171] *AHDR*, p. 232.

[172] *AHDR*, chapter 4, and pp. 232-233.

[173] *AHDR*, chapters 6-7, and pp. 232-233.

[174] *ACIA*, pp. 1000-1001, 1004.

[175] *ACIA*, pp. 1000-1001, 1004.

[176] Parkinson, Alan J. *The Arctic Human Health Initiative*. Washington DC: U.S. Department of Health and Human Services, 2006; Brubaker, Michael, James Berner, Raj Chavan, and John Warren. "Climate Change and Health Effects in Northwest Alaska." *Global Health Action* 4 (October 18, 2011).

[177] U.S. Government Accountability Office (GAO), *Alaska Native Villages: Villages Affected by Flooding and Erosion Have Difficulty Qualifying for Federal Assistance*, GAO-04-895T, June 29, 2004, p. i, http://www.gao.gov/new.items/ d04895t.pdf. See also, Government Accountability Office, *Alaska Native Villages: Most Are Affected by Flooding and Erosion, but Few Qualify for Federal Assistance*, GAO-04-142, December 12, 2003, http://www.gao.gov/new.items/ d04142.pdf.

[178] U.S. Army Corps of Engineers. Alaska Baseline Erosion Assessment (BEA), March 2009.

[179] GAO, *Alaska Native Villages: Limited Progress Has Been Made on Relocating Villages Threatened by Flooding and Erosion*, June 3, 2009.

[180] Alaska Federation of Natives, Human Resources Committee, *2010 Federal Priorities* (Anchorage: Alaska Federation of Natives, 2010), pp. 22-23, available at http://www.nativefederation.org/documents/ 2010_AFN_Federal_Priorities.pdf. See also, Government Accountability Office, *Alaska Native Villages: Limited Progress Has Been Made on Relocating Villages Threatened by Flooding and Erosion*, GAO-09-551, June 3, 2009, http://www.gao.gov/new.items/d09551.pdf. For a more detailed anecdote, see a 2012 interview with Brice Eningowuk, Shishmaref Erosion and Relocation Committee member at Shearer, Christine. "Climate Crisis: Alaskan Village Shishmaref Sinking Into the Sea." *Imagined Magazine*, July 8, 2012, http://imaginedmag.com/2012/07/climate-crisis- alaskan-village-shishmaref-sinking-into-the-sea/.

[181] Press Office of Senator Lisa Murkowski. "Murkowski Urges Greater Tribal Consultation from Administration; Senator Spotlights 'Monumental' Climate Struggles Faced by Alaska Natvies, Coastal Communities", July 19, 2012.

[182] *ACIA*, pp. 1001, 1004.

[183] Aqqaluk Lynge, Chair, Inuit Circumpolar Council, quoted in Stephanie McFeeters, "Lynge talks future of Inuit people," *The Dartmouth*, February 8, 2012.

[184] See, for example, "Health: Increased Bacterial Loads in Potable Water Could Have Significant Health Effects on Indigenous People From the Arctic to Uganda, Says Vanier Scholar." National Aboriginal Health Organization (NAHO), February 17, 2012. http://www.naho.ca/blog/2012/02/17/health-increased-bacterial-loads-in-potable-water-

could-have-significant-health-effects-on-indigenous-people-from-the-arctic-to-uganda-says-vanier-scholar/; or, Kallenborn et al. *Combined Effects of Selected Pollutants and Climate Change in the Arctic Environment.* Oslo, Norway: Arctic Monitoring and Assessment Programme (AMAP), Arctic Council, 2011.

[185] *AMAP Assessment 2009: Human Health in the Arctic*, ed. Simon J. Wilson and Carolyn Symon (Oslo: Arctic Monitoring and Assessment Programme, 2009), pp. 4-6, 143.

[186] Brubaker, Michael, James Berner, Raj Chavan, and John Warren. "Climate Change and Health Effects in Northwest Alaska." *Global Health Action* 4 (October 18, 2011); John Warren, "Climate change could affect human health," *Mukluk Telegraph*, January/February 2005, pp. 5-6.

[187] John Warren, "Climate change could affect human health," *Mukluk Telegraph*, January/February 2005, pp. 5-6.

[188] *AHDR*, p. 235.

[189] See http://www.arctic-council.org/. The six organizations are the Aleut International Association, Arctic Athabaskan Council, Gwich'in Council International, Inuit Circumpolar Council, RAIPON (Russian Association of Indigenous Peoples of the North), and Saami Council.

[190] United Nations, Declaration on the Rights of Indigenous Peoples, Resolution A/RES/61/295, at http://social.un.org/index/IndigenousPeoples/DeclarationontheRightsofIndigenousPeoples.aspx.

[191] See http://www.indigenoussummit.com/servlet/content/home.html.

[192] K. Galloway-McLean et al., *Report of the Indigenous Peoples' Global Summit on Climate Change: 20-24 April 2009, Anchorage, Alaska* (Darwin, Australia: United Nations University—Traditional Knowledge Initiative, 2009), pp. 5-7; available at http://www.indigenoussummit.com/servlet/content/home.html.

[193] This section prepared by Ronald O'Rourke, Specialist in Naval Affairs, Foreign Affairs, Defense, and Trade Division. It is adapted from CRS Report RL34391, *Coast Guard Polar Icebreaker Modernization: Background and Issues for Congress*, by Ronald O'Rourke.

[194] The Coast Guard's polar icebreakers also support U.S. scientific research in the Antarctic.

[195] *United States Coast Guard Arctic Strategy*, Washington, May 2013, p. 35, at http://www.uscg.mil/seniorleadership/ DOCS/CG_Arctic_Strategy.pdf.

[196] Source: Coast Guard briefing to CRS, June 14, 2013.

[197] *United States Coast Guard High Latitude Region Mission Analysis Capstone Summary*, July 2010, pp. 12-13, 15.

[198] Source: Transcript of hearing.

[199] This section prepared by Ronald O'Rourke, Specialist in Naval Affairs, Foreign Affairs, Defense, and Trade Division.

[200] Dan Elliott, "NORAD Chiefs Eye Busier Arctic," *Fairbanks Daily News-Miner*, May 7, 2011.

[201] For a report assessing certain emergency scenarios in the Arctic, including search and rescue scenarios, see *Opening the Arctic Seas, Envisioning Disasters and Framing Solutions*, Coastal Response and Research Center, University of New Hampshire, report of January 2009, based on conference held March 18-20, 2008, at Durham, New Hampshire.

[202] Alister Doyle, "Arctic Nations Must Urgently Improve Rescue Services—Canadian Experts," Reuters (http://in.reuters.com), May 13, 2013.

[203] Cid Standifer, "Coast Guard Comandant: Service Still Committed To Eight NSCs," Inside the Navy, January 17, 2011.

[204] Jennifer Scholtes, "Papp: Coast Guard Arctic Plan Must Look Beyond Icebreakers," *CQ Homeland Security*, March 1, 2011.

[205] Source: Transcript of hearing.

[206] Source: State Department website, http://www.state.gov/g/oes/ocns/opa/arc/c29382.htm, July 7, 2011.

[207] The guidelines, as stated in Article 7, are as follows:

(a) search and rescue operations conducted pursuant to this Agreement in the territory of a Party shall be carried out consistent with the laws and regulations of that Party;
(b) if a search and rescue agency and/or RCC of a Party receives information that any person is, or appears to be, in distress, that Party shall take urgent steps to ensure that the necessary assistance is provided;
(c) any Party having reason to believe that a person, a vessel or other craft or aircraft is in a state of emergency in the area of another Party as set forth in paragraph 2 of the Annex shall forward as soon as possible all available information to the Party or Parties concerned;
(d) the search and rescue agency and/or RCC of a Party that has received information concerning a situation provided for in subparagraph (b) of this paragraph may request assistance from the other Parties;
(e) the Party to whom a request for assistance is submitted shall promptly decide on and inform the requesting Party whether or not it is in a position to render the assistance requested and shall promptly indicate the scope and the terms of the assistance that can be rendered;
(f) the Parties shall ensure that assistance be provided to any person in distress. They shall do so regardless of the nationality or status of such a person or the circumstances in which that person is found; and
(g) a Party shall promptly provide all relevant information regarding the search and rescue of any person to the consular or diplomatic authorities concerned.

208 Article 20 identifies the government of Canada as the depository for the agreement.

209 Source: Text of final version of agreement made ready for signing and dated April 21, 2011, http://arctic- council.org/filearchive/Arctic_SAR_Agreement_EN_FINAL_for_signature_ 21-Apr-2011.pdf

210 Source: State Department website,http://www.state.gov/r/pa/prs/ps/2011/05/163285.htm, July 7, 2011.

211 This section prepared by Carl Ek, Specialist in International Relations, Foreign Affairs, Defense, and Trade Division.

212 "Arctic Sea Ice Continues to Decline, Hits Near Record Low," *IB Times*, October 4, 2011.

213 "China Pushes for Arctic Foothold – From a Thousand Miles Away," Christian Science Monitor Global News Blog, March 7, 2013.

214 "China, India Enter Heating-up Arctic Race," *Asia Times,* January 25, 2012; "Aston in Far North to Reach EU Policy on Arctic," *Agence France Presse,* March 8, 2012; "EU Application to Arctic Council Postponed," North Norway European Office, May 18, 2011 http://www.northnorway.org/nyheter/847-eu-application-to-arctic-council- postponed See also: "Arctic Governance: Balancing Challenges and Development," European Parliament Regional Briefing 2012, DG EXPO/B/PolDep/Note/2012_136 June, 2012

215 Conference on the Arctic Ocean, May 26, 2008, Arctic Council website: http://arctic-council.org/article/2008/5/ conference_on_the_arctic_ocean.

216 Conference In Ilulissat, Greenland: Landmark Political Declaration on the Future of the Arctic, Edited December 11, 2008, website of the Danish Foreign Ministry, http://www.um.dk/en.

217 The Council members are the "Arctic 5" coastal States plus Sweden, Finland, and Iceland. Six indigenous Arctic peoples organizations are permanent participants. Several other countries, including Spain, France, Germany, the Netherlands, Poland, and the UK, have permanent observer status. At the May 2013 Council summit, six additional countries, including China and India, were granted observer status. For additional information, see the Arctic Council's website: http://www.arctic-council.org/index.php/en/. "Document On Cooperation In Arctic Search and Rescue Efforts Could Be Ready In April 2011," *Interfax: Russia & CIS Military Newswire*, February 26, 2010.

218 "States Set Rules on Exploiting Arctic Wealth," *Agence France Presse*, May 12, 2011.

219 "Secretary Clinton Signs the Arctic Search and Rescue Agreement with Other Arctic Nations," U.S. Department of State website, May 12, 2011. "Warming Arctic Opens Way to Competition For Resources," *Washington Post*, May 16, 2011.

[220] "Clinton in Arctic to See Impact of Climate Change," *Agence France Presse,* June 2, 2012.

[221] The European Union (EU) had applied for observer status, but its request was put on hold pending the resolution of a dispute with Canada and indigenous peoples regarding commercial sealing.

[222] "The Arctic Council: Underpinning Stability in the Arctic," The Arctic Institute/Centre for Circumpolar Security Studies; March 26, 2013, http://www.thearcticinstitute.org/2013/03/the-arctic-council-underpinning.html

[223] "Nations Ponder How to Handle Busier, More Polluted Arctic," *Climatewire,* May 13, 2013, http://www.eenews.net/climatewire/stories/1059981040

[224] "The Arctic: Special Report," *The Economist,* June 16, 2012, p. 11.

[225] During an October 2011 conference on Arctic shipping, President Putin enthused that "the Arctic is the shortcut between the largest markets of Europe and the Asia-Pacific region.... It is an excellent opportunity to optimize costs." "Warming Revives Dream of Sea Route in Russian Arctic," *New York Times,* October 17, 2011.

[226] "Foreign Ministers of Russia and Canada Agreed To Go To the Ice Without the Third Parties," *WPS: Defense & Security,* September 20, 2010. "Russia Will Staunchly Defend its Interests in the Arctic Region – Putin," *Interfax.* June 30, 2011. "Russia to Deploy Troops to Defend Interests in Arctic," *Wall Street Journal,* July 2, 2011; "Russia Lets China Into Arctic Rush as Energy Giants Embrace," *Bloomberg.com,* March 25, 2013; "Russia, U.S., Norway to Hold Joint Naval Drills in August," *RIA Novosti,* May 24, 2012 http://en.rian.ru/world/20120524/173654802.html "Norway Hails Northern Eagle as Bridge-builder," *Defense New;* August 24, 2012; "China Eyes Arctic Options in Energy, Transport," *CNN Wire.* May 16, 2013.

[227] It was recently noted that China in December 2012 "offered Greenland a $5 billion investment package that included rights to mine iron, gold, nickel, and rare earth metals ...." See "Denmark, Norway Focus on Arctic Security," *Defense News,* March 11, 2013.

[228] "What Is China's Arctic Game Plan?" *The Atlantic,* May 2013.

[229] "China Plans First Commercial Trip through Arctic Shortcut in 2013," *Reuters.* March 12, 2013. "China Pushes for Arctic Foothold—From a Thousand Miles Away," Christian Science Monitor Global News Blog, March 7, 2013.

[230] "The Dragon Eyes the Top of the World," David Curtis Wright, Naval War College, China Maritime Studies Institute, Number 8, August 2011.

[231] "China Defines Itself as a 'Near-Arctic State,' Says SIPRI," Swedish International Peace Research Institute, May 10, 2012. "China's Arctic Ambitions," *Stratfor,* June 18, 2012. "China: Icebreaking in the Arctic," *The ISN Blog,* May 4, 2012; "Ice Station Dragon: China's Strategic Arctic Interest," *Defense News,* May 16, 2011.

[232] "China Knocks on Iceland's Door," *New York Times,* March 12, 2013; "China Signs Free Trade Agreement with Iceland as it Looks for Foothold in the Arctic," *The Washington Post,* April 16, 2013.

[233] "Chinese Icebreaker Concludes Arctic Expedition," *Xinhua News Agency,* September 27, 2012.

[234] National Strategy for the Arctic Region, The White House. May 2013, http://www.whitehouse.gov/sites/default/ files/docs/nat_arctic_strategy.pdf

[235] Heather Conley, "The Colder War: U.S., Russia and Others Are Vying for Control of Santa's Back Yard," *Washington Post,* December 23, 2011.

[236] "Welcome To a New Era of Arctic Security," Rob Huebert, Canadian Defense and Foreign Affairs Institute, August, 2010.

[237] See, for example, "Canada Vows 'Firm' Defence of Arctic Border," *Canwest News Service,* November 24, 2009. "Danes With Dogs To Join Military Sovereignty Patrol Of Canadian Arctic," *The Canadian Press,* March 3, 2010. "Norway Plans One New Arctic Base, Perhaps More," *Defense News,* March 8, 2010.

[238] "The Arctic: Special Report," *The Economist,* June 16, 2012, p. 10.

[239] "Hands Across the Melting Ice," by James F. Collins, ross A. Virginia, and Kenneth Yalowitz, *New York Times,* May 13, 2013. "Oil Executives Tune Out the Call of the Wild Arctic," *Reuters,* May 31, 2013.

[240] "A New Security Architecture for the Arctic: An American Perspective," Center for Strategic and International Studies (CSIS), January 2012, p. 30.

[241] "Russia, Norway Sign Border Deal For Arctic Energy," *Reuters*, September 15, 2010. "NATO Chief Cautions Against Division Over Arctic," *Canwest News Service*, January 29, 2009. "NATO Proposes Arctic Cooperation With Russia," Reuters, January 29, 2009. Speech by NATO Secretary General Jaap de Hoop Scheffer On Security Prospects In the High North, January 29, 2009, http://www.nato.int/docu/speech/2009/s090129a.html. "NATO Chief Wars Of Climate Change Security Risks," *Agence France Presse*, October 1, 2009. "Rogozin Says He Won't Discuss Cooperation In Arctic With NATO," *Interfax: Russia & CIS General Newswire,* March 27, 2009.

[242] "NATO Tests Its Forces in Arctic," February 16, 2012, Atlantic Council web page: http://www. acus.org/natosource/nato-tests-its-forces-arctic "NATO, Russia Stage Arctic War Games," April 25, 2012 Atlantic Council web page: http://www. acus.org/print/68922 See also Cold Response 2012, Norwegian Armed Forces web page: http://mil.no/excercises/ coldresponse2012/pages/default.aspx "Øvelse Cold Response er en norskledet, internasjonal øvelse som foregår i store deler av Nord-Norge," Norwegian Ministry of Defense, January 21, 2013, http://forsvaret.no/aktuelt/ovelser/Sider/Cold-Response.aspx.

[243] "Security in the High North: NATO's Role," NATO Parliamentary Assembly draft report, 071 PCTR 13 E, March 28, 2013.

[244] "NATO Won't Up Presence in the Arctic: Chief," *Agence France Presse,* May 8, 2013. "NATO Rejects Direct Arctic Presence," *Defense News,* May 27, 2013.

[245] "Russia Will Protect Interests in Arctic: Official," *Agence France Presse,* June 10, 2009. "Russia Opposes Arms Race In Arctic Region—Diplomat," *ITAR-TASS World Service,* July 21, 2009.

[246] "China Prepares For an Ice-free Arctic," Linda Jakobson, *SIPRI Insights On Peace and Security*, No. 2010/2, March 2010. "Admiral Urges Government to Stake Claim in the Arctic," *South China Morning Post*, March 6, 2010. "Norway Wants to Block China from Arctic Council Over 2010 Nobel Peace Prize Row," *Agence France Presse*, January 25, 2012. "Clinton in Arctic to See Impact of Climate Change," *Agence France Presse,* June 2, 2012.

[247] "China Pours Cash into Melting Arctic in Bid to Win Influence," *RIA-Oreanda News* [Russia], April 2, 2013.

[248] "The Arctic Council: Underpinning Stability in the Arctic," the Arctic Institute/Centre for Circumpolar Security Studies, March 26, 2013. http://www.thearcticinstitute.org/2013/03/ the-arctic-council-underpinning.html.

[249] "A New Security Architecture for the Arctic: An American Perspective," Center for Strategic and International Studies (CSIS), January 2012, p. 37.

[250] "Arctic Nations Meet to Discuss Communication, Maritime Domain Awareness Strategy," EUCOM website, August 30, 2012, http://www.eucom.mil/blog-post/24109/arctic-nations-meet-to-discuss-communication-maritime- domain-awareness-strategy.

[251] "General Natynczyk and Fellow Northern Chiefs of Defence Discuss Shared Arctic Interests," News release, Canadian National Defense web page, April 13, 2012 http://www.forces. gc.ca/site/news-nouvelles/news-nouvelles- eng.asp?id=4151 "Northcom Strives to Promote Safe, Secure Arctic," U.S. Department of Defense web page, December 17, 2012, http://www.defense.gov/news/newsarticle.aspx?id=118814.

[252] This section prepared by Ronald O'Rourke, Specialist in Naval Affairs, Foreign Affairs, Defense, and Trade Division.

[253] See, for example, John Vandiver, "NATO Commander Sees Arctic Seabed As Cooperative Zone," *Mideast Stars and Stripes*, October 10, 2009; Tom Coghlan, "Nato Commander Warns Of Conflict With Russia In Arctic Circle," *London Times*, October 3, 2009; Gerard

O'Dwyer, "Danish Report: Conflicts Coming Over Arctic," *Defense News*, September 28, 2009: 18; Gerard O'Dwyer, "Russia Warns Denmark Over Arctic Arms Race," *Defense News*, August 3, 2009: 13; Gerrard Cowen, "Russia and NATO Look to Co-operation In Arctic," *Jane's Defence Weekly*, April 8, 2009: 12; David Scutro, "Thawing Arctic Seas May Bring Security Risk," *NavyTimes.com*, January 29, 2009.

[254] Department of Defense, *Quadrennial Defense Review Report*, February 2010, p. 86.

[255] Department of Defense, *Quadrennial Defense Review Report*, February 2010, p. 19.

[256] Department of Defense, *Quadrennial Defense Review Report*, February 2010, p. 57.

[257] Department of Defense, *Quadrennial Defense Review Report*, February 2010, p. 62.

[258] Jim Garamone, "Unified Command Plan Reflects Arctic's Importance," *American Forces Press Service*, April 7, 2011.

[259] The direction was contained on page 337 of H.Rept. 111-491 of May 21, 2010, the House Armed Services Committee's report on H.R. 5136, the FY2011 National Defense Authorization Act.

[260] Department of Defense, *Report to Congress on Arctic Operations and the Northwest Passage*, OUSD (Policy), May 2011. pp 2-4 (executive summary). See also Christopher J. Castelli, "Report: 'Significant' Gap Undermines DOD Tracking Of Ships In Arctic," *Inside the Pentagon*, June 16, 2011.

[261] The direction was contained on page 291 of H.Rept. 112-78 of May 17, 2011, the House Armed Services Committee's report on H.R. 1540, the FY2012 National Defense Authorization Act.

[262] Government Accountability Office, *Arctic Capabilities[:]DOD Addressed Many Specified Reporting Elements in Its 2011 Arctic Report but Should Take Steps to Meet Near- and Long-term Needs*, GAO-12-180, January 2012, Summary page and pp. 12-18.

[263] NSPD 66/HSPD 25, Section III B.

[264] David Grove, "Arctic Melt: Reopening a Naval Frontier?" *U.S. Naval Institute Proceedings*, February 2009: 16-21.

[265] David W. Titley and Courtney C. St. John, "Arctic Security Considerations and the U.S. Navy's Roadmap for the Arctic," *Naval War College Review*, Spring 2010: 41.

[266] David W. Titley and Courtney C. St. John, "Arctic Security Considerations and the U.S. Navy's Roadmap for the Arctic," *Naval War College Review*, Spring 2010: 42.

[267] Source: Text of remarks of Chief of Naval Operations Admiral Gary Roughead at Active in the Arctic Seminar, June 16, 2011.

[268] Source: Text of remarks of Chief of Naval Operations Admiral Gary Roughead at the 4th Symposium on the Impacts of an Ice-Diminishing Arctic on Naval & Maritime Operations, June 20, 2011. See also Christopher J. Castelli, "Roughead: International Approach Needed To Fix Arctic-Awareness Gap," *Inside the Navy*, June 27, 2011.

[269] Nicole Klauss, "War Games Find U.S. Navy Ability Lacking In Arctic," *Fairbanks Daily News-Miner*, April 28, 2012.

[270] Memorandum for Distribution dated November 10, 2009, from Admiral J. W. Greenert, Vice Chief of Naval Operations, on the subject of the Navy Arctic Roadmap, accessed July 24, 2013 at http://www.navy.mil/navydata/documents/USN_artic_roadmap.pdf.

[271] U.S. Navy, U.S. Navy Arctic Roadmap, Washington, 2009, 29 pp. (October 2009, sponsored by Task Force Climate Change [and] Oceanographer of the Navy).

[272] Statement of Admiral Jonathan Greenert, U.S. Navy, Chief of Naval Operations, Before the House Armed Services Committee on FY 2014 Department of the Navy Posture, April 16, 2013, p. 21. See also Olga Belogolova, "Navy Updating Roadmap, Strategy For Arctic Region This Year," *Inside the Navy*, July 22, 2013.

[273] Department of the Navy, *Arctic Environmental Assessment and Outlook Report*, August 2011, 25 pp,, accessed July 24, 2013, at: http://greenfleet.dodlive.mil/files/2011/08/U.S.-Navy-Arctic-Environmental-Assessment.pdf.

[274] The POM is an internal DOD document that guides the development of DOD's annual budget submissions to Congress.

[275] Department of the Navy, *Arctic Environmental Assessment and Outlook Report*, August 2011, p. v.

[276] Nicole Klauss, "War Games Find U.S. Navy Ability Lacking In Acrtic," *Fairbanks Daily News-Miner*, April 28, 2012.

[277] For additional discussion, see "Polar Icebreaking".

[278] *United States Coast Guard High Latitude Region Mission Analysis Capstone Summary*, July 2010, pp. 10-11, 13-15.

[279] *United States Coast Guard Arctic Strategy*, Washington, May 2013, 47 pp.; accessed May 24, 2013, at http://www.uscg.mil/seniorleadership/DOCS/CG_Arctic_Strategy.pdf.

[280] *United States Coast Guard Arctic Strategy*, Washington, May 2013, pp. 9-10. Emphasis as in original.

[281] Antonieta Rico, "Papp: No Plans For More Coasties in Arctic," *NavyTimes.com*, June 1, 2013.

[282] H.R. 3619/P.L. 111-281 of October 15, 2010 is the Coast Guard Authorization Act of 2010. Section 307 concerned implementation of the Arctic Marine Shipping Assessment. Subsection 307(b) stated that "To carry out the purpose of this section, the Secretary of the department in which the Coast Guard is operating is encouraged to enter into negotiations through the International Maritime Organization to conclude and execute agreements to promote coordinated action among the United States, Russia, Canada, Iceland, Norway, and Denmark and other seafaring and Arctic nations to ensure, in the Arctic--(1) placement and maintenance of aids to navigation; (2) appropriate marine safety, tug, and salvage capabilities; (3) oil spill prevention and response capability; (4) maritime domain awareness, including long-range vessel tracking; and (5) search and rescue.

[283] Christopher J. Castelli, "DOD, Canada Aim TO Develop Arctic Policy Options By Early 2010," *Inside the Navy*, December 7, 2009.

[284] Christopher J. Castelli, "American, Canadian Defense Officials Tighten Ties On Arctic Issues," *Inside the Navy*, May 3, 2010.

[285] NORAD news release, "NORAD, USNORTHCOM and Canadian Joint Operations Commmand Commander Sign Cooperative U.S./Canada Documents," December 11, 2012, accessed December 21, 2012, at http://www.norad.mil/ News/2012/121112a.html.

[286] Gerard O'Dwyer, "Arctic Nations Set Cooperation Guidelines," *DefenseNews.com*, June 27, 2013. See also Gerard O'Dwyer, "Greenland Meeting Highlights Arctic's Growing Importance," *DefenseNews.com*, June 2, 2013.

[287] CRS communication with State Department official, October 8, 2010.

[288] On this question, a January 19, 2009, press article about the new directive stated:

The new policy directive covers several key areas, including national security, energy exploration and the environment, but it does not specify whether any should take precedence over others.

That led Jeremy Rabkin, a professor at George Mason University Law School, to comment: "It's really a list of all the things we're concerned about; that's not policy. I don't see anything here that helps you decide what gets priority."

(Juliet Eilperin and Spencer S. Hsu, "White House Directive Guides Policy On Arctic," *Washington Post*, January 19, 2009: 2.)

[289] Heather A. Conley et al, *The New Foreign Policy Frontier[:] U.S. Interests and Actors in the Arctic*, Washington, Center for Strategic & International Studies, March 2013, pp. 21-22.

[290] Title II of P.L. 98-373 is the National Critical Materials Act of 1984.

[291] Source for text: http://fas.org/irp/offdocs/nspd/nspd-66.htm. The text is also available at http://www.nsf.gov/od/opp/opp_advisory/briefings/may2009/nspd66_hspd25.pdf.

INDEX

D

E

F

G

H

I

J

K

L

M

N

O

P

T

U

V

W